HANDBOOK OF MACHINING AND METALWORKING CALCULATIONS

HANDBOOK OF MACHINING AND METALWORKING CALCULATIONS

Ronald A. Walsh

McGRAW-HILL
New York San Francisco Washington, D.C. Auckland Bogotá
Caracas Lisbon London Madrid Mexico City Milan
Montreal New Delhi San Juan Singapore
Sydney Tokyo Toronto

Cataloging-in-Publication Data is on file with the Library of Congress

McGraw-Hill
A Division of The McGraw-Hill Companies

Copyright © 2001 by The McGraw-Hill Companies, Inc. All rights reserved. Printed in the United States of America. Except as permitted under the United States Copyright Act of 1976, no part of this publication may be reproduced or distributed in any form or by any means, or stored in a data base or retrieval system, without the prior written permission of the publisher.

1 2 3 4 5 6 7 8 9 0   DOC/DOC   0 9 8 7 6 5 4 3 2 1 0

ISBN 0-07-136066-2

The sponsoring editor for this book was Scott Grillo, the editing supervisor was M.R. Carey, and the production supervisor was Pamela A. Pelton. It was set in Times Ten Roman by North Market Street Graphics.

Printed and bound by R. R. Donnelley & Sons Company.

McGraw-Hill books are available at special quantity discounts to use as premiums and sales promotions, or for use in corporate training programs. For more information, please write to the Director of Special Sales, McGraw-Hill, 2 Penn Plaza, New York, NY 10121-2298. Or contact your local bookstore.

> Information contained in this work has been obtained by The McGraw-Hill Companies, Inc. ("McGraw-Hill") from sources believed to be reliable. However, neither McGraw-Hill nor its authors guarantee the accuracy or completeness of any information published herein, and neither McGraw-Hill nor its authors shall be responsible for any errors, omissions, or damages arising out of use of this information. This work is published with the understanding that McGraw-Hill and its authors are supplying information but are not attempting to render engineering or other professional services. If such services are required, the assistance of an appropriate professional should be sought.

 This book was printed on recycled, acid-free paper containing a minimum of 50% recycled, de-inked fiber

CONTENTS

Preface ix

Chapter 1. Mathematics for Machinists and Metalworkers   **1.1**

1.1 Geometric Principles—Plane Geometry / *1.1*
1.2 Basic Algebra / *1.7*
    1.2.1 Algebraic Procedures / *1.7*
    1.2.2 Transposing Equations (Simple and Complex) / *1.9*
1.3 Plane Trigonometry / *1.11*
    1.3.1 Trigonometric Laws / *1.13*
    1.3.2 Sample Problems Using Trigonometry / *1.21*
1.4 Modern Pocket Calculator Procedures / *1.28*
    1.4.1 Types of Calculators / *1.28*
    1.4.2 Modern Calculator Techniques / *1.29*
    1.4.3 Pocket Calculator Bracketing Procedures / *1.31*
1.5 Angle Conversions—Degrees and Radians / *1.32*
1.6 Powers-of-Ten Notation / *1.34*
1.7 Percentage Calculations / *1.35*
1.8 Temperature Systems and Conversions / *1.36*
1.9 Decimal Equivalents and Millimeters / *1.37*
1.10 Small Weight Equivalents: U.S. Customary (Grains and Ounces) Versus Metric (Grams) / *1.38*
1.11 Mathematical Signs and Symbols / *1.39*

Chapter 2. Mensuration of Plane and Solid Figures   **2.1**

2.1 Mensuration / *2.1*
2.2 Properties of the Circle / *2.10*

Chapter 3. Layout Procedures for Geometric Figures   **3.1**

3.1 Geometric Constructions / *3.1*

Chapter 4. Measurement and Calculation Procedures for Machinists   **4.1**

4.1 Sine Bar and Sine Plate Calculations / *4.1*
4.2 Solutions to Problems in Machining and Metalworking / *4.6*
4.3 Calculations for Specific Machining Problems (Tool Advance, Tapers, Notches and Plugs, Diameters, Radii, and Dovetails) / *4.15*
4.4 Finding Complex Angles for Machined Surfaces / *4.54*

Chapter 5. Formulas and Calculations for Machining Operations   5.1

5.1 Turning Operations / *5.1*
5.2 Threading and Thread Systems / *5.12*
5.3 Milling / *5.22*
5.4 Drilling and Spade Drilling / *5.38*
5.5 Reaming / *5.61*
5.6 Broaching / *5.63*
5.7 Vertical Boring and Jig Boring / *5.66*
5.8 Bolt Circles (BCs) and Hole Coordinate Calculations / *5.67*

Chapter 6. Formulas for Sheet Metal Layout and Fabrication   6.1

6.1 Sheet Metal Flat-Pattern Development and Bending / *6.8*
6.2 Sheet Metal Developments, Transitions, and Angled Corner Flange Notching / *6.14*
6.3 Punching and Blanking Pressures and Loads / *6.32*
6.4 Shear Strengths of Various Materials / *6.32*
6.5 Tooling Requirements for Sheet Metal Parts—Limitations / *6.36*

Chapter 7. Gear and Sprocket Calculations   7.1

7.1 Involute Function Calculations / *7.1*
7.2 Gearing Formulas—Spur, Helical, Miter/Bevel, and Worm Gears / *7.4*
7.3 Sprockets—Geometry and Dimensioning / *7.15*

Chapter 8. Ratchets and Cam Geometry   8.1

8.1 Ratchets and Ratchet Gearing / *8.1*
8.2 Methods for Laying Out Ratchet Gear Systems / *8.3*
  8.2.1 External-Tooth Ratchet Wheels / *8.3*
  8.2.2 Internal-Tooth Ratchet Wheels / *8.4*
  8.2.3 Calculating the Pitch and Face of Ratchet-Wheel Teeth / *8.5*
8.3 Cam Layout and Calculations / *8.6*

Chapter 9. Bolts, Screws, and Thread Calculations   9.1

9.1 Pullout Calculations and Bolt Clamp Loads / *9.1*
9.2 Measuring and Calculating Pitch Diameters of Threads / *9.5*
9.3 Thread Data (UN and Metric) and Torque Requirements (Grades 2, 5, and 8 U.S. Standard 60° V) / *9.13*

Chapter 10. Spring Calculations—Die and Standard Types   10.1

10.1 Helical Compression Spring Calculations / *10.5*
  10.1.1 Round Wire / *10.5*
  10.1.2 Square Wire / *10.6*
  10.1.3 Rectangular Wire / *10.6*
  10.1.4 Solid Height of Compression Springs / *10.6*
10.2 Helical Extension Springs (Close Wound) / *10.8*

CONTENTS vii

10.3 Spring Energy Content of Compression and Extension Springs / *10.8*
10.4 Torsion Springs / *10.11*
    10.4.1 Round Wire / *10.11*
    10.4.2 Square Wire / *10.12*
    10.4.3 Rectangular Wire / *10.13*
    10.4.4 Symbols, Diameter Reduction, and Energy Content / *10.13*
10.5 Flat Springs / *10.14*
10.6 Spring Materials and Properties / *10.16*
10.7 Elastomer Springs / *10.22*
10.8 Bending and Torsional Stresses in Ends of Extension Springs / *10.23*
10.9 Specifying Springs, Spring Drawings, and Typical Problems and Solutions / *10.24*

Chapter 11. Mechanisms, Linkage Geometry, and Calculations   11.1

11.1 Mathematics of the External Geneva Mechanism / *11.1*
11.2 Mathematics of the Internal Geneva Mechanism / *11.3*
11.3 Standard Mechanisms / *11.5*
11.4 Clamping Mechanisms and Calculation Procedures / *11.9*
11.5 Linkages—Simple and Complex / *11.17*

Chapter 12. Classes of Fit for Machined Parts—Calculations   12.1

12.1 Calculating Basic Fit Classes (Practical Method) / *12.1*
12.2 U.S. Customary and Metric (ISO) Fit Classes and Calculations / *12.5*
12.3 Calculating Pressures, Stresses, and Forces Due to Interference Fits, Force Fits, and Shrink Fits / *12.9*

Index   I.1

PREFACE

This handbook contains most of the basic and advanced calculation procedures required for machining and metalworking applications. These calculation procedures should be performed on a modern pocket calculator in order to save time and reduce or eliminate errors while improving accuracy. Correct bracketing procedures are required when entering equations into the pocket calculator, and it is for this reason that I recommend the selection of a calculator that shows all entered data on the calculator display and that can be scrolled. That type of calculator will allow you to scroll or review the entered equation and check for proper bracketing sequences, prior to pressing "*ENTER*" or =. If the bracketing sequences of an entered equation are incorrect, the calculator will indicate "Syntax error," or give an incorrect solution to the problem. Examples of proper bracketing for entering equations in the pocket calculator are shown in Chap. 1 and in Chap. 11, where the complex four-bar linkage is analyzed and explained.

This book is written in a user-friendly format, so that the mathematical equations and examples shown for solutions to machining and metalworking problems are not only highly useful and relatively easy to use, but are also practical and efficient. This book covers metalworking mathematics problems, from the simple to the highly complex, in a manner that should be valuable to all readers.

It should be understood that these mathematical procedures are applicable for:

- Master machinists
- Machinists
- Tool designers and toolmakers
- Metalworkers in various fields
- Mechanical designers
- Tool engineering personnel
- CNC machining programmers
- The gunsmithing trade
- Students in technical teaching facilities

R.A. Walsh

HANDBOOK OF MACHINING AND METALWORKING CALCULATIONS

CHAPTER 1
MATHEMATICS FOR MACHINISTS AND METALWORKERS

This chapter covers all the basic and special mathematical procedures of value to the modern machinist and metalworker. Geometry and plane trigonometry are of prime importance, as are the basic algebraic manipulations. Solutions to many basic and complex machining and metalworking operations would be difficult or impossible without the use of these branches of mathematics. In this chapter and other subsections of the handbook, all the basic and important aspects of these branches of mathematics will be covered in detail. Examples of typical machining and metalworking problems and their solutions are presented throughout this handbook.

1.1 GEOMETRIC PRINCIPLES— PLANE GEOMETRY

In any triangle, angle A + angle B + angle C = 180°, and angle A = 180° − (angle A + angle B), and so on (see Fig. 1.1). If three sides of one triangle are proportional to the corresponding sides of another triangle, the triangles are similar. Also, if $a:b:c = a':b':c'$, then angle A = angle A', angle B = angle B', angle C = angle C', and $a/a' = b/b' = c/c'$. Conversely, if the angles of one triangle are equal to the respective angles of another triangle, the triangles are similar and their sides proportional; thus if angle A = angle A', angle B = angle B', and angle C = angle C', then $a:b:c = a':b':c'$ and $a/a' = b/b' = c/c'$ (see Fig. 1.2).

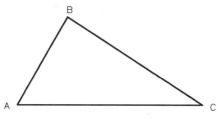

FIGURE 1.1 Triangle.

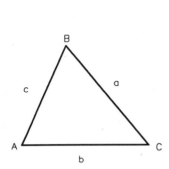 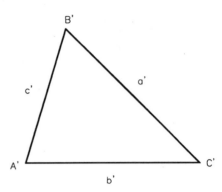

FIGURE 1.2 Similar triangles.

Isosceles triangle (see Fig. 1.3). If side c = side b, then angle C = angle B.

Equilateral triangle (see Fig. 1.4). If side a = side b = side c, angles A, B, and C are equal (60°).

Right triangle (see Fig. 1.5). $c^2 = a^2 + b^2$ and $c = (a^2 + b^2)^{1/2}$ when angle $C = 90°$. Therefore, $a = (c^2 - b^2)^{1/2}$ and $b = (c^2 - a^2)^{1/2}$. This relationship in all right-angle triangles is called the *Pythagorean theorem*.

Exterior angle of a triangle (see Fig. 1.6). Angle C = angle A + angle B.

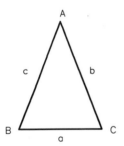 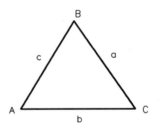

FIGURE 1.3 Isosceles triangle.

FIGURE 1.4 Equilateral triangle.

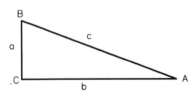 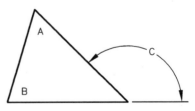

FIGURE 1.5 Right-angled triangle.

FIGURE 1.6 Exterior angle of a triangle.

Intersecting straight lines (see Fig. 1.7). Angle A = angle A', and angle B = angle B'.

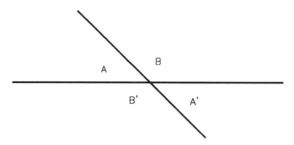

FIGURE 1.7 Intersecting straight lines.

Two parallel lines intersected by a straight line (see Fig. 1.8). Alternate interior and exterior angles are equal: angle A = angle A'; angle B = angle B'.

Any four-sided geometric figure (see Fig. 1.9). The sum of all interior angles = 360°; angle A + angle B + angle C + angle D = 360°.

A line tangent to a point on a circle is at 90°, or normal, to a radial line drawn to the tangent point (see Fig. 1.10).

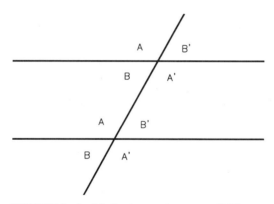

FIGURE 1.8 Straight line intersecting two parallel lines.

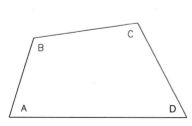

FIGURE 1.9 Quadrilateral (four-sided figure).

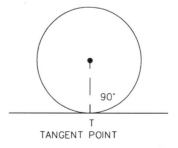

FIGURE 1.10 Tangent at a point on a circle.

Two circles' common point of tangency is intersected by a line drawn between their centers (see Fig. 1.11).

Side $a = a'$; angle A = angle A' (see Fig. 1.12).

Angle A = ½ angle B (see Fig. 1.13).

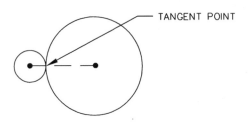

FIGURE 1.11 Common point of tangency.

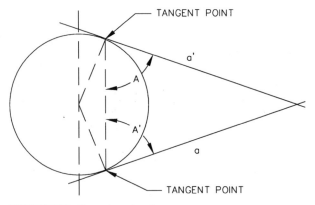

FIGURE 1.12 Tangents and angles.

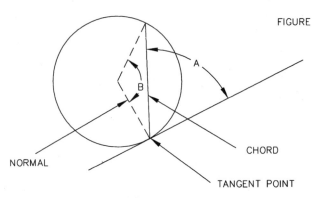

FIGURE 1.13 Half-angle (A).

Angle A = angle B = angle C. All perimeter angles of a chord are equal (see Fig. 1.14).
Angle B = ½ angle A (see Fig. 1.15).
$a^2 = bc$ (see Fig. 1.16).
All perimeter angles in a circle, drawn from the diameter, are 90° (see Fig. 1.17).
Arc lengths are proportional to internal angles (see Fig. 1.18). Angle A:angle B = a:b. Thus, if angle A = 89°, angle B = 30°, and arc a = 2.15 units of length, arc b would be calculated as

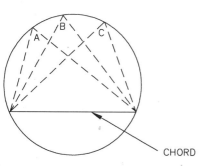
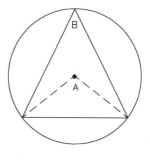

FIGURE 1.14 Perimeter angles of a chord.   **FIGURE 1.15** Half-angle (B).

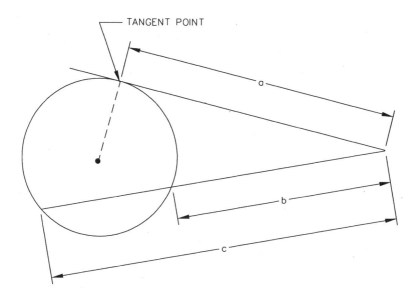

FIGURE 1.16 Line and circle relationship ($a^2 = bc$).

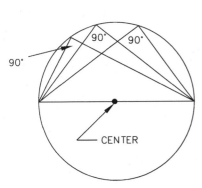

FIGURE 1.17 90° perimeter angles.

FIGURE 1.18 Proportional arcs and angles.

$$\frac{\text{Angle } A}{\text{Angle } B} = \frac{a}{b}$$

$$\frac{89}{30} = \frac{2.15}{b}$$

$$89b = 30 \times 2.15$$

$$b = \frac{64.5}{89}$$

$$b = 0.7247 \text{ units of length}$$

NOTE. The angles may be given in decimal degrees or radians, consistently.

Circumferences are proportional to their respective radii (see Fig. 1.19). $C:C' = r:R$, and areas are proportional to the squares of the respective radii.

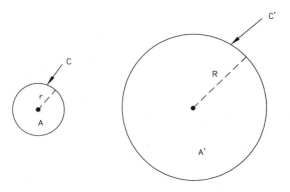

FIGURE 1.19 Circumference and radii proportionality.

1.2 BASIC ALGEBRA

1.2.1 Algebraic Procedures

Solving a Typical Algebraic Equation. An algebraic equation is solved by substituting the numerical values assigned to the variables which are denoted by letters, and then finding the unknown value, using algebraic procedures.

EXAMPLE

$$L = 2C + 1.57(D + d) + \frac{(D-d)^2}{4C} \quad \text{(belt-length equation)}$$

If $C = 16$, $D = 5.56$, and $d = 3.12$ (the variables), solve for L (substituting the values of the variables into the equation):

$$L = 2(16) + 1.57(5.56 + 3.12) + \frac{(5.56 - 3.12)^2}{4(16)}$$

$$= 32 + 1.57(8.68) + \frac{(2.44)^2}{64}$$

$$= 32 + 13.628 + \frac{5.954}{64}$$

$$= 32 + 13.628 + 0.093$$

$$= 45.721$$

Most of the equations shown in this handbook are solved in a similar manner, that is, by substituting known values for the variables in the equations and solving for the unknown quantity using standard algebraic and trigonometric rules and procedures.

Ratios and Proportions. If $a/b = c/d$, then

$$\frac{a+b}{b} = \frac{c+d}{d}; \quad \frac{a-b}{b} = \frac{c-d}{d} \quad \text{and} \quad \frac{a-b}{a+b} = \frac{c-d}{c+d}$$

Quadratic Equations. Any quadratic equation may be reduced to the form

$$ax^2 + bx + c = 0$$

The two roots, x_1 and x_2, equal

$$\frac{-b \pm \sqrt{b^2 - 4ac}}{2a} \quad (x_1 \text{ use } +; x_2 \text{ use } -)$$

When a, b, and c are real, if $b^2 - 4ac$ is positive, the roots are real and unequal. If $b^2 - 4ac$ is zero, the roots are real and equal. If $b^2 - 4ac$ is negative, the roots are imaginary and unequal.

Radicals

$$a^0 = 1$$

$$(\sqrt[n]{a})^n = a$$

$$\sqrt[n]{a^n} = a$$

$$\sqrt[n]{ab} = n\sqrt[n]{a} \times n\sqrt[n]{b}$$

$$\sqrt[n]{\frac{a}{b}} = \sqrt[n]{a} \div \sqrt[n]{b}$$

$$\sqrt[n]{a^x} = a^{x/n} \quad \text{hence } \sqrt[3]{7^2} = 7^{2/3}$$

$$\sqrt[n]{a} = a^{1/n} \quad \text{hence } \sqrt{3} = 3^{1/2}$$

$$a^{-n} = \frac{1}{a^n}$$

Factorial. 5! is termed *5 factorial* and is equivalent to

$$5 \times 4 \times 3 \times 2 \times 1 = 120$$

$$9! = 9 \times 8 \times 7 \times 6 \times 5 \times 4 \times 3 \times 2 \times 1 = 362{,}880$$

Logarithms. The logarithm of a number N to base a is the exponent power to which a must be raised to obtain N. Thus $N = a^x$ and $x = \log_a N$. Also $\log_a 1 = 0$ and $\log_a a = 1$.

Other relationships follow:

$$\log_a MN = \log_a M + \log_a N$$

$$\log_a \frac{M}{N} = \log_a M - \log_a N$$

$$\log_a N^k = k \log_a N$$

$$\log_a \sqrt[n]{N} = \frac{1}{n} \log_a N$$

$$\log_b a = \frac{1}{\log_a b} \quad \text{let } N = a$$

Base 10 logarithms are referred to as *common logarithms* or *Briggs logarithms*, after their inventor.

Base e logarithms (where $e = 2.71828$) are designated as *natural, hyperbolic,* or *Naperian logarithms,* the last label referring to their inventor. The base of the natural logarithm system is defined by the infinite series

MATHEMATICS FOR MACHINISTS AND METALWORKERS

$$e = 1 + \frac{1}{1} + \frac{1}{2!} + \frac{1}{3!} + \frac{1}{4!} + \frac{1}{5!} + \cdots = \lim_{n \to \infty} \left(1 + \frac{1}{n}\right)^n$$

$e = 2.71828\ldots$

If a and b are any two bases, then

$$\log_a N = (\log_a b)(\log_b N)$$

or

$$\log_b N = \frac{\log_a N}{\log_a b}$$

$$\log_{10} N = \frac{\log_e N}{2.30261} = 0.43429 \log_e N$$

$$\log_e N = \frac{\log_{10} N}{0.43429} = 2.30261 \log_{10} N$$

Simply multiply the natural log by 0.43429 (a modulus) to obtain the equivalent common log.

Similarly, multiply the common log by 2.30261 to obtain the equivalent natural log. (Accuracy is to four decimal places for both cases.)

1.2.2 Transposing Equations (Simple and Complex)

Transposing an Equation. We may solve for any one unknown if all other variables are known. The given equation is:

$$R = \frac{Gd^4}{8ND^3}$$

An equation with five variables, shown in terms of R. Solving for G:

$Gd^4 = R8ND^3$ (cross-multiplied)

$$G = \frac{8RND^3}{d^4}$$ (divide both sides by d^4)

Solving for d:

$$Gd^4 = 8RND^3$$

$$d^4 = \frac{8RND^3}{G}$$

$$d = \sqrt[4]{\frac{8RND^3}{G}}$$

Solving for D:

$$Gd^4 = 8RND^3$$

$$D^3 = \frac{Gd^4}{8RN}$$

$$D = \sqrt[3]{\frac{Gd^4}{8RN}}$$

Solve for N using the same transposition procedures shown before.

NOTE. When a complex equation needs to be transposed, shop personnel can contact their engineering or tool engineering departments, where the MathCad program is usually available.

Transposing Equations using MathCad (Complex Equations). The transposition of basic algebraic equations has many uses in the solution of machining and metalworking problems. Transposing a complex equation requires considerable skill in mathematics. To simplify this procedure, the use of MathCad is invaluable. As an example, a basic equation involving trigonometric functions is shown here, in its original and transposed forms. The transpositions are done using symbolic methods, with degrees or radians for the angular values.

Basic Equation

$$L = X + d \cdot \left[\left(\tan\left(\frac{90 - \alpha}{2}\right)\right) + 1\right]$$

Transposed Equations (Angles in Degrees)

$$\text{Solve}, \alpha \rightarrow 90 + 2 \cdot \operatorname{atan}\left[\frac{(-L + X + d)}{d}\right] \quad \text{Solved for } \alpha$$

$$\text{Solve}, X \rightarrow L + d \cdot \tan\left(-45 + \frac{1}{2} \cdot \alpha\right) - d \quad \text{Solved for } X$$

$$\text{Solve}, d \rightarrow \frac{(-L + X)}{\left(\tan\left(-45 + \frac{1}{2} \cdot \alpha\right) - 1\right)} \quad \text{Solved for } d$$

NOTE. The angular values are expressed in degrees.

Basic Equation

$$L = X + d \cdot \left[\left(\tan\left(\frac{\frac{\pi}{2} - \alpha}{2}\right)\right) + 1\right]$$

Transposed Equations (Angles in Radians)

$$\text{Solve, } \alpha \rightarrow \frac{3}{2} \cdot \pi - 2 \cdot \text{acot}\left[\frac{(-L + X + d)}{d}\right] \quad \text{Solved for } \alpha$$

$$\text{Solve, } X \rightarrow L - d \cdot \cot\left(\frac{1}{4} \cdot \pi + \frac{1}{2} \cdot \alpha\right) - d \quad \text{Solved for } X$$

$$\text{Solve, } d \rightarrow \frac{-(-L + X)}{\left(\cot\left(\frac{1}{4} \cdot \pi + \frac{1}{2} \cdot \alpha\right) + 1\right)} \quad \text{Solved for } d$$

NOTE. The angular values are expressed in radians, i.e., 90 degrees = $\pi/2$ radians; 2π radians = 360°; π radians = 180°.

1.3 PLANE TRIGONOMETRY

There are six trigonometric functions: sine, cosine, tangent, cotangent, secant, and cosecant. The relationships of the trigonometric functions are shown in Fig. 1.20. Trigonometric functions shown for angle A (right-angled triangle) include

$$\sin A = a/c \text{ (sine)}$$
$$\cos A = b/c \text{ (cosine)}$$
$$\tan A = a/b \text{ (tangent)}$$
$$\cot A = b/a \text{ (cotangent)}$$
$$\sec A = c/b \text{ (secant)}$$
$$\csc A = c/a \text{ (cosecant)}$$

For angle B, the functions would become (see Fig. 1.20)

$$\sin B = b/c \text{ (sine)}$$
$$\cos B = a/c \text{ (cosine)}$$

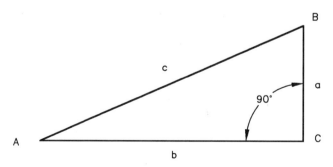

FIGURE 1.20 Right-angled triangle.

CHAPTER ONE

$\tan B = b/a$ (tangent)

$\cot B = a/b$ (cotangent)

$\sec B = c/a$ (secant)

$\csc B = c/b$ (cosecant)

As can be seen from the preceding, the sine of a given angle is always the side opposite the given angle divided by the hypotenuse of the triangle. The cosine is always the side adjacent to the given angle divided by the hypotenuse, and the tangent is always the side opposite the given angle divided by the side adjacent to the angle. These relationships *must* be remembered at all times when performing trigonometric operations. Also:

$$\sin A = \frac{1}{\csc A}$$

$$\cos A = \frac{1}{\sec A}$$

$$\tan A = \frac{1}{\cot A}$$

This reflects the important fact that the cosecant, secant, and cotangent are the reciprocals of the sine, cosine, and tangent, respectively. This fact also *must* be remembered when performing trigonometric operations.

Signs and Limits of the Trigonometric Functions. The following coordinate chart shows the sign of the function in each quadrant and its numerical limits. As an example, the sine of any angle between 0 and 90° will always be positive, and its numerical value will range between 0 and 1, while the cosine of any angle between 90 and 180° will always be negative, and its numerical value will range between 0 and 1. Each quadrant contains 90°; thus the fourth quadrant ranges between 270 and 360°.

Quadrant II		Quadrant I
$(1-0)+\sin$		$\sin+(0-1)$
$(0-1)-\cos$	y	$\cos+(1-0)$
$(\infty-0)-\tan$		$\tan+(0-\infty)$
$(0-\infty)-\cot$		$\cot+(\infty-0)$
$(\infty-1)-\sec$		$\sec+(1-\infty)$
$(1-\infty)+\csc$		$\csc+(\infty-1)$
Quadrant III	0	**Quadrant IV**
$(0-1)-\sin$		$\sin-(1-0)$
$(1-0)-\cos$		$\cos+(0-1)$
$(0-\infty)+\tan$		$\tan-(\infty-0)$
$(\infty-0)+\cot$		$\cot-(0-\infty)$
$(1-\infty)-\sec$		$\sec+(\infty-1)$
$(\infty-1)-\csc$	y'	$\csc-(1-\infty)$

(x' ——— x axis through center)

1.3.1 Trigonometric Laws

The trigonometric laws show the relationships between the sides and angles of non-right-angle triangles or oblique triangles and allow us to calculate the unknown parts of the triangle when certain values are known. Refer to Fig. 1.21 for illustrations of the trigonometric laws that follow.

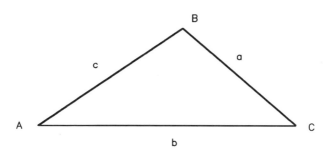

FIGURE 1.21 Oblique triangle.

The Law of Sines. See Fig. 1.21.

$$\frac{a}{\sin A} = \frac{b}{\sin B} = \frac{c}{\sin C}$$

And,
$$\frac{a}{b} = \frac{\sin A}{\sin B} \qquad \frac{b}{c} = \frac{\sin B}{\sin C} \qquad \frac{a}{c} = \frac{\sin A}{\sin C}$$

Also, $a \times \sin B = b \times \sin A$; $b \times \sin C = c \times \sin B$, etc.

The Law of Cosines. See Fig. 1.21.

$$\left. \begin{array}{l} a^2 = b^2 + c^2 - 2bc \cos A \\ b^2 = a^2 + c^2 - 2ac \cos B \\ c^2 = a^2 + b^2 - 2ab \cos C \end{array} \right\} \text{May be transposed as required}$$

The Law of Tangents. See Fig. 1.21.

$$\frac{a+b}{a-b} = \frac{\tan \dfrac{A+B}{2}}{\tan \dfrac{A-B}{2}}$$

With the preceding laws, the trigonometric functions for right-angled triangles, the Pythagorean theorem, and the following triangle solution chart, it will be possible to find the solution to any plane triangle problem, provided the correct parts are specified.

The Solution of Triangles

In right-angled triangles	To solve
Known: Any two sides	Use the Pythagorean theorem to solve unknown side; then use the trigonometric functions to solve the two unknown angles. The third angle is 90°.
Known: Any one side and either one angle that is not 90°	Use trigonometric functions to solve the two unknown sides. The third angle is 180° − sum of two known angles.
Known: Three angles and no sides (*all* triangles)	Cannot be solved because there are an infinite number of triangles which satisfy three known internal angles.
Known: Three sides	Use trigonometric functions to solve the two unknown angles.

In oblique triangles	To solve
Known: Two sides and any one of two nonincluded angles	Use the law of sines to solve the second unknown angle. The third angle is 180° − sum of two known angles. Then find the other sides using the law of sines or the law of tangents.
Known: Two sides and the included angle	Use the law of cosines for one side and the law of sines for the two angles.
Known: Two angles and any one side	Use the law of sines to solve the other sides or the law of tangents. The third angle is 180° − sum of two known angles.
Known: Three sides	Use the law of cosines to solve two of the unknown angles. The third angle is 180° − sum of two known angles.
Known: One angle and one side (non right triangle)	Cannot be solved except under certain conditions. If the triangle is equilateral or isosceles, it may be solved if the known angle is opposite the known side.

Finding Heights of Non-Right-Angled Triangles. The height x shown in Figs. 1.22 and 1.23 is found from

$$x = b \, \frac{\sin A \sin C}{\sin (A + C)} = \frac{b}{\cot A + \cot C} \quad \text{(for Fig. 1.22)}$$

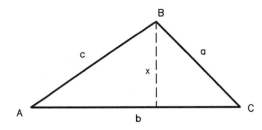

FIGURE 1.22 Height of triangle x.

MATHEMATICS FOR MACHINISTS AND METALWORKERS　　1.15

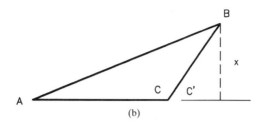

FIGURE 1.23　Height of triangle x.

$$x = b\,\frac{\sin A \sin C}{\sin(C' - A)} = \frac{b}{\cot A - \cot C'} \quad \text{(for Fig. 1.23)}$$

Areas of Triangles.　See Fig. 1.24a and b.

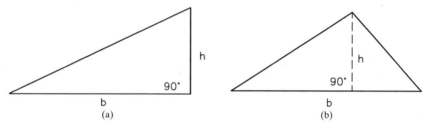

FIGURE 1.24　Triangles: (*a*) right triangle; (*b*) oblique triangle.

$$A = \frac{1}{2}bh$$

The area when the three sides are known (see Fig. 1.25) (this holds true for any triangle):

$$A = \sqrt{s(s-a)(s-b)(s-c)}$$

where
$$s = \frac{a+b+c}{2}$$

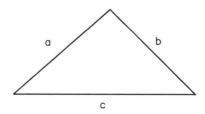

FIGURE 1.25　Triangle.

The Pythagorean Theorem. For right-angled triangles:

$$c^2 = a^2 + b^2$$
$$b^2 = c^2 - a^2$$
$$a^2 = c^2 - b^2$$

NOTE. Side c is the hypotenuse.

Practical Solutions to Triangles. The preceding sections concerning the basic trigonometric functions and trigonometric laws, together with the triangle solution chart, will allow you to solve all plane triangles, both their parts and areas. Whenever you solve a triangle, the question always arises, "Is the solution correct?" In the engineering office, the triangle could be drawn to scale using AutoCad and its angles and sides measured, but in the shop this cannot be done with accuracy. In machining, gearing, and tool engineering problems, the triangle must be solved with great accuracy and its solution verified.

To verify or check the solution of triangles, we have the Mollweide equation, which involves all parts of the triangle. By using this classic equation, we know if the solution to any given triangle is correct or if it has been calculated correctly.

The Mollweide Equation

$$\frac{a-b}{c} = \frac{\sin\left(\frac{A-B}{2}\right)}{\cos\left(\frac{C}{2}\right)}$$

Substitute the calculated values of all sides and angles into the Mollweide equation and see if the equation balances algebraically. Use of the Mollweide equation will be shown in a later section. Note that the angles must be specified in decimal degrees when using this equation.

Converting Angles to Decimal Degrees. Angles given in degrees, minutes, and seconds must be converted to decimal degrees prior to finding the trigonometric functions of the angle on modern hand-held calculators.

Converting Degrees, Minutes, and Seconds to Decimal Degrees
Procedure. Convert $26°41'26''$ to decimal degrees.

Degrees = 26.000000   in decimal degrees

Minutes = 41/60 = 0.683333   in decimal parts of a degree

Seconds = 26/3600 = 0.007222   in decimal parts of a degree

The angle in decimal degrees is then

$$26.000000 + 0.683333 + 0.007222 = 26.690555°$$

MATHEMATICS FOR MACHINISTS AND METALWORKERS   **1.17**

Converting Decimal Degrees to Degrees, Minutes, and Seconds
Procedure. Convert 56.5675 decimal degrees to degrees, minutes, and seconds.

Degrees = 56 degrees

Minutes = 0.5675 × 60 = 34.05 = 34 minutes

Seconds = 0.05 (minutes) × 60 = 3 seconds

The answer, therefore, is 56°34′3″.

Summary of Trigonometric Procedures for Triangles. There are four possible cases in the solution of oblique triangles:

Case 1. Given one side and two angles: a, A, B
Case 2. Given two sides and the angle opposite them: a, b, A or B
Case 3. Given two sides and their included angle: a, b, C
Case 4. Given the three sides: a, b, c

All oblique (non-right-angle) triangles can be solved by use of natural trigonometric functions: the law of sines, the law of cosines, and the angle formula, angle A + angle B + angle C = 180°. This may be done in the following manner:

Case 1. Given $a, A,$ and B, angle C may be found from the angle formula; then sides b and c may be found by using the law of sines twice.

Case 2. Given $a, b,$ and A, angle B may be found by the law of sines, angle C from the angle formula, and side c by the law of sines again.

Case 3. Given $a, b,$ and C, side c may be found by the law of cosines, and angles A and B may be found by the law of sines used twice; or angle A from the law of sines and angle B from the angle formula.

Case 4. Given $a, b,$ and c, the angles may all be found by the law of cosines; or angle A may be found from the law of cosines, and angles B and C from the law of sines; or angle A from the law of cosines, angle B from the law of sines, and angle C from the angle formula.

In all cases, the solutions may be checked with the Mollweide equation.

NOTE. Case 2 is called the *ambiguous case*, in which there may be one solution, two solutions, or *no* solution, given $a, b,$ and A.

- If angle $A < 90°$ and $a < b \sin A$, there is *no* solution.
- If angle $A < 90°$ and $a = b \sin A$, there is one solution—a right triangle.
- If angle $A < 90°$ and $b > a > b \sin A$, there are two solutions—oblique triangles.
- If angle $A < 90°$ and $a \geqq b$, there is one solution—an oblique triangle.
- If angle $A < 90°$ and $a \leqq b$, there is *no* solution.
- If angle $A > 90°$ and $a > b$, there is one solution—an oblique triangle.

Mollweide Equation Variations. There are two forms for the Mollweide equation:

$$\frac{a+b}{c} = \frac{\cos\left(\frac{A-B}{2}\right)}{\sin\left(\frac{C}{2}\right)}$$

$$\frac{a-b}{c} = \frac{\sin\left(\frac{A-B}{2}\right)}{\cos\left(\frac{C}{2}\right)}$$

Use either form for checking triangles.

The Accuracy of Calculated Angles

Required accuracy of the angle	Significant figures required in distances
10 minutes	3
1 minute	4
10 seconds	5
1 second	6

Special Half-Angle Formulas. In case 4 triangles where only the three sides a, b, and c are known, the sets of half-angle formulas shown here may be used to find the angles:

$$\sin\frac{A}{2} = \sqrt{\frac{(s-b)(s-c)}{bc}} \qquad \cos\frac{B}{2} = \sqrt{\frac{s(s-b)}{ac}}$$

$$\sin\frac{B}{2} = \sqrt{\frac{(s-c)(s-a)}{ca}} \qquad \cos\frac{C}{2} = \sqrt{\frac{s(s-c)}{ab}}$$

$$\sin\frac{C}{2} = \sqrt{\frac{(s-a)(s-b)}{ab}} \qquad \tan\frac{A}{2} = \sqrt{\frac{(s-b)(s-c)}{s(s-a)}}$$

$$\cos\frac{A}{2} = \sqrt{\frac{s(s-a)}{bc}} \qquad \tan\frac{B}{2} = \sqrt{\frac{(s-c)(s-a)}{s(s-b)}}$$

$$\tan\frac{C}{2} = \sqrt{\frac{(s-a)(s-b)}{s(s-c)}}$$

where $\qquad s = \sqrt{\dfrac{a+b+c}{2}}$

Additional Relations of the Trigonometric Functions

$$\sin x = \cos(90° - x) = \sin(180° - x)$$

$$\cos x = \sin(90° - x) = -\cos(180° - x)$$

$$\tan x = \cot(90° - x) = -\tan(180° - x)$$

$$\cot x = \tan(90° - x) = -\cot(180° - x)$$

$$\csc x = \cot\frac{x}{2} - \cot x$$

Functions of Half-Angles

$$\sin\frac{1}{2}x = \pm\sqrt{\frac{1 - \cos x}{2}}$$

$$\cos\frac{1}{2}x = \pm\sqrt{\frac{1 + \cos x}{2}}$$

$$\tan\frac{1}{2}x = \pm\sqrt{\frac{1 - \cos x}{1 + \cos x}} = \frac{1 - \cos x}{\sin x} = \frac{\sin x}{1 + \cos x}$$

NOTE. The sign before the radical depends on the quadrant in which $x/2$ falls. See functions in the four quadrants chart in the text.

Functions of Multiple Angles

$$\sin 2x = 2 \sin x \cos x$$

$$\cos 2x = \cos^2 x - \sin^2 x = 2\cos^2 x - 1 = 1 - 2\sin^2 x$$

$$\tan 2x = \frac{2 \tan x}{1 - \tan^2 x}$$

$$\cot 2x = \frac{\cot^2 x - 1}{2 \cot x}$$

Functions of Sums of Angles

$$\sin(x \pm y) = \sin x \cos y \pm \cos x \sin y$$

$$\cos(x \pm y) = \cos x \cos y \mp \sin x \sin y$$

$$\tan(x + y) = \frac{\tan x \pm \tan y}{1 \mp \tan x \tan y}$$

Miscellaneous Relations

$$\tan x \pm \tan y = \frac{\pm \sin(x \pm y)}{\sin x \sin y}$$

$$\frac{1 + \tan x}{1 - \tan x} = \tan(45° + x)$$

$$\frac{\cot x + 1}{\cot x - 1} = \cot(45° - x)$$

$$\frac{\sin x + \sin y}{\sin x - \sin y} = \frac{\tan \frac{1}{2}(x+y)}{\tan \frac{1}{2}(x-y)}$$

Relations Between Sides and Angles of Any Plane Triangle

$$a = b \cos C + c \cos B$$

$$\cos A = \frac{b^2 + c^2 - a^2}{2bc}$$

$$\tan\left(\frac{A-B}{2}\right) = \frac{a-b}{a+b} \cot \frac{C}{2}$$

$$\sin A = \frac{2}{bc} \sqrt{s(s-a)(s-b)(s-c)}$$

where $s = \frac{1}{2}(a + b + c)$

$$r = \sqrt{\frac{(s-a)(s-b)(s-c)}{s}}$$

$$\sin \frac{A}{2} = \sqrt{\frac{(s-b)(s-c)}{bc}}$$

$$\cos \frac{A}{2} = \sqrt{\frac{s(s-a)}{bc}}$$

$$\tan \frac{A}{2} = \sqrt{\frac{(s-b)(s-c)}{s(s-a)}} = \frac{r}{s-a}$$

MATHEMATICS FOR MACHINISTS AND METALWORKERS   1.21

$$\frac{a+b}{a-B} = \frac{\sin A + \sin B}{\sin A - \sin B} = \frac{\tan\frac{1}{2}(A+B)}{\tan\frac{1}{2}(A-B)} = \frac{\cot\frac{1}{2}C}{\tan\frac{1}{2}(A-B)}$$

Trigonometric Functions Reduced to the First Quadrant.   See Fig. 1.26.

FIGURE 1.26   Trigonometric functions reduced to first quadrant.

	If ∠ α in degrees, is between:		
	90–180	180–270	270–360
	First subtract:		
	α − 90	α − 180	α − 270
	Then:		
sin α	= +cos (α − 90)	= −sin (α − 180)	= −cos (α − 270)
cos α	= −sin (α − 90)	= −cos (α − 180)	= +sin (α − 270)
tan α	= −cot (α − 90)	= +tan (α − 180)	= −cot (α − 270)
cot α	= −tan (α − 90)	= +cot (α − 180)	= −tan (α − 270)
sec α	= −csc (α − 90)	= −sec (α − 180)	= +csc (α − 270)
csc α	= +sec (α − 90)	= −csc (α − 180)	= −sec (α − 270)

1.3.2   Sample Problems Using Trigonometry

Samples of Solutions to Triangles

Solving Right-Angled Triangles by Trigonometry.   *Required:* Any one side and angle *A* or angle *B* (see Fig. 1.27). Solve for side *a:*

$$\sin A = \frac{a}{c}$$

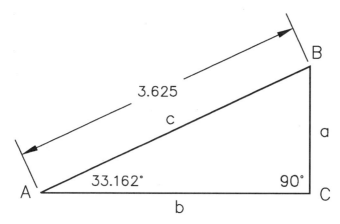

FIGURE 1.27   Solve the triangle.

$$\sin 33.162° = \frac{a}{3.625}$$

$$a = 3.625 \times \sin 33.162°$$

$$= 3.625 \times 0.5470$$

$$= 1.9829$$

Solve for side b:

$$\cos A = \frac{b}{c}$$

$$\cos 33.162° = \frac{b}{3.625}$$

$$b = 3.625 \times \cos 33.162°$$

$$b = 3.625 \times 0.8371$$

$$b = 3.0345$$

Then

$$\text{angle } B = 180° - (\text{angle } A + 90°)$$

$$= 180° - 123.162°$$

$$= 56.838°$$

We now know sides a, b, and c and angles A, B, and C.

Solving Non-Right-Angled Triangles Using the Trigonometric Laws. Solve the triangle in Fig. 1.28. *Given:* Two angles and one side:

$$A = 45°$$

$$B = 109°$$

$$a = 3.250$$

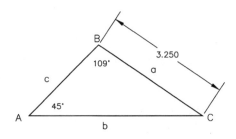

FIGURE 1.28 Solve the triangle.

First, find angle C:

$$\text{Angle } C = 180° - (\text{angle } A + \text{angle } B)$$
$$= 180° - (45° + 109°)$$
$$= 180° - 154°$$
$$= 26°$$

Second, find side b by the law of sines:

$$\frac{a}{\sin A} = \frac{b}{\sin B}$$

$$\frac{3.250}{0.7071} = \frac{b}{0.9455}$$

Therefore,

$$b = \frac{3.250 \times 0.9455}{0.7071}$$

$$= 4.3457$$

Third, find side c by the law of sines:

$$\frac{a}{\sin A} = \frac{c}{\sin C}$$

$$\frac{3.250}{0.7071} = \frac{c}{0.4384}$$

Therefore,

$$c = \frac{3.250 \times 0.4384}{0.7071}$$

$$= 2.0150$$

The solution to this triangle has been calculated as $a = 3.250$, $b = 4.3457$, $c = 2.0150$, angle $A = 45°$, angle $B = 109°$, and angle $C = 26°$.

We now use the Mollweide equation to check the calculated answer by substituting the parts into the equation and checking for a balance, which signifies equality and the correct solution.

$$\frac{a-b}{c} = \frac{\sin\left(\frac{A-B}{2}\right)}{\cos\left(\frac{C}{2}\right)}$$

$$\frac{3.250 - 4.3457}{2.0150} = \frac{\sin\left(\frac{45 - 109}{2}\right)}{\cos\left(\frac{26}{2}\right)}$$

$$\frac{-1.0957}{2.0150} = \frac{\sin(-32°)}{\cos 13°} \quad \text{(Find sin −32° and cos 13° on a calculator.)}$$

$$\frac{-1.0957}{2.0150} = \frac{-0.5299}{0.9744} \quad \text{(Divide both sides.)}$$

$$-0.5438 = -0.5438 \quad \text{(Cross-multiplying will also show an equality.)}$$

This equality shows that the calculated solution to the triangle shown in Fig. 1.28 is correct.

Solve the triangle in Fig. 1.29. *Given:* Two sides and one angle:

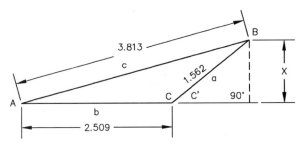

FIGURE 1.29 Solve the triangle.

$$\text{Angle } A = 16°$$
$$a = 1.562$$
$$b = 2.509$$

First, find angle B from the law of sines:

$$\frac{a}{\sin A} = \frac{b}{\sin B}$$

$$\frac{1.562}{\sin 16} = \frac{2.509}{\sin B}$$

$$\frac{1.562}{0.2756} = \frac{2.509}{\sin B}$$

$$1.562 \cdot \sin B = 0.6915 \quad \text{(by cross-multiplication)}$$

$$\sin B = \frac{0.6915}{1.562}$$

$$\sin B = 0.4427$$

$$\text{arccos } 0.4427 = 26.276° = \text{angle } B$$

Second, find angle C:

$$\text{Angle } C = 180° - (\text{angle } A + \text{angle } B)$$

$$= 180° - 42.276°$$

$$= 137.724°$$

Third, find side c from the law of sines:

$$\frac{a}{\sin A} = \frac{c}{\sin C}$$

$$\frac{1.562}{0.2756} = \frac{c}{0.6727}$$

$$0.2756c = 1.0508$$

$$c = 3.813$$

We may now find the altitude or height x of this triangle (see Fig. 1.29). Refer to Fig. 1.23 and text for the following equation for x.

$$x = b \frac{\sin A \sin C}{\sin (C' - A)} \quad (\text{where angle } C' = 180° - 137.724° = 42.276° \text{ in Fig. 1.29})$$

$$= 2.509 \times \frac{0.2756 \times 0.6727}{\sin (42.276 - 16)}$$

$$= 2.509 \times \frac{0.1854}{0.4427}$$

$$= 2.509 \times 0.4188$$

$$= 1.051$$

This height x also can be found from the sine function of angle C', when side a is known, as shown here:

$$\sin C' = \frac{x}{1.562}$$

$$x = 1.562 \sin C' = 1.562 \times 0.6727 = 1.051$$

Both methods yield the same numerical solution: 1.051. Also, the preceding solution to the triangle shown in Fig. 1.29 is correct because it will balance the Mollweide equation.

Solve the triangle in Fig. 1.30. *Given:* Three sides and no angles. According to the preceding triangle solution chart, solving this triangle requires use of the law of cosines. Proceed as follows. First, solve for any angle (we will take angle C first):

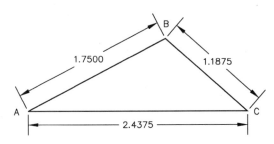

FIGURE 1.30   Solve the triangle.

$$c^2 = a^2 + b^2 - 2ab \cos C$$

$$(1.7500)^2 = (1.1875)^2 + (2.4375)^2 - 2(1.1875 \times 2.4375) \cos C$$

$$3.0625 = 1.4102 + 5.9414 - 5.7891 \cos C$$

$$5.7891 \cos C = 1.4102 + 5.9414 - 3.0625$$

$$\cos C = \frac{4.2891}{5.7891}$$

$$\cos C = 0.7409$$

arccos $0.7409 = 42.192° =$ angle C (the angle whose cosine is 0.7409)

Second, by the law of cosines, find angle B:

$$b^2 = a^2 + c^2 - 2ac \cos B$$

$$(2.4375)^2 = (1.1875)^2 + (1.7500)^2 - 2(1.1875 \times 1.7500) \cos B$$

$$5.9414 = 1.4102 + 3.0625 - 4.1563 \cos B$$

$$4.1563 \cos B = 1.4102 + 3.0625 - 5.9414$$

$$\cos B = \frac{-1.4687}{4.1563}$$

$$\cos B = -0.3534$$

arccos $-0.3534 = 110.695° =$ angle B (the angle whose cosine is −0.3534)

Then, angle A is found from

$$\text{angle } A = 180 - (42.192 + 110.695)$$
$$= 180 - 152.887$$
$$= 27.113°$$

The solution to the triangle shown in Fig. 1.30 is therefore $a = 1.1875$, $b = 2.4375$, $c = 1.7500$ (given), angle $A = 27.113°$, angle $B = 110.695°$, and angle $C = 42.192°$ (calculated). This also may be checked using the Mollweide equation.

Proof of the Mollweide Equation. From the Pythagorean theorem it is known and can be proved that any triangle with sides equal to 3 and 4 and a hypotenuse of 5 will be a perfect right-angled triangle. Multiples of the numbers 3, 4, and 5 also produce perfect right-angled triangles, such as 6, 8, and 10, etc. ($c^2 = a^2 + b^2$).

If you solve the 3, 4, and 5 proportioned triangle for the internal angles and then substitute the sides and angles into the Mollweide equation, it will balance, indicating that the solution is valid mathematically.

A note on use of the Mollweide equation when checking triangles: If the Mollweide equation does not balance,

- The solution to the triangle is incorrect.
- The solution is not accurate.
- The Mollweide equation was incorrectly calculated.
- The triangle is not "closed," or the sum of the internal angles does not equal 180°.

Natural Trigonometric Functions. There are no tables of natural trigonometric functions or logarithms in this handbook. This is due to the widespread availability of the electronic digital calculator. You may find these numerical values quicker and more accurately than any table can provide. See Sec. 1.4 for calculator uses and techniques applicable to machining and metalworking practices.

The natural trigonometric functions for sine, cosine, and tangent may be calculated using the following infinite-series equations. The cotangent, secant, and cosecant functions are merely the numerical reciprocals of the tangent, cosine, and sine functions, respectively.

$$\frac{1}{\text{tangent}} = \text{cotangent}$$

$$\frac{1}{\text{cosine}} = \text{secant}$$

$$\frac{1}{\text{sine}} = \text{cosecant}$$

Calculating the Natural Trigonometric Functions. Infinite series for the sine (angle x must be given in radians):

$$\sin x = x - \frac{x^3}{3!} + \frac{x^5}{5!} - \frac{x^7}{7!} + \frac{x^9}{9!} - \frac{x^{11}}{11!} + \cdots$$

Infinite series for the cosine (angle x must be given in radians):

$$\cos x = 1 - \frac{x^2}{2!} + \frac{x^4}{4!} - \frac{x^6}{6!} + \frac{x^8}{8!} - \frac{x^{10}}{10!} + \cdots$$

The natural tangent may now be found from the sine and cosine series using the equality

$$\tan x = \frac{\sin x \text{ (series)}}{\cos x \text{ (series)}}$$

1.4 MODERN POCKET CALCULATOR PROCEDURES

1.4.1 Types of Calculators

The modern hand-held or pocket digital electronic calculator is an invaluable tool to the machinist and metalworker. Many cumbersome tables such as natural trigonometric functions, powers and roots, sine bar tables, involute functions, and logarithmic tables are not included in this handbook because of the ready availability, simplicity, speed, and great accuracy of these devices.

Typical multifunction pocket calculators are shown in Fig. 1.31. This type of device will be used to illustrate the calculator methods shown in Sec. 1.4.2 following.

FIGURE 1.31 Typical standard pocket calculators.

MATHEMATICS FOR MACHINISTS AND METALWORKERS   1.29

The advent of the latest generation of hand-held programmable calculators—including the Texas Instruments TI-85 and Hewlett Packard HP-48G (see Fig. 1.32)—has made possible many formerly difficult or nearly impossible engineering computations. Both instruments have enormous capabilities in solving complex general mathematical problems. See Sec. 11.5 for a complete explanation for applying these calculators to the important and useful four-bar linkage mechanism, based on use of the standard Freudenstein equation.

FIGURE 1.32 Programmable calculators with complex equation-solving ability and other advanced features. HP-48G on the right and the TI-85 on the left.

Some of the newer machines also do not rely on battery power, since they have a built-in high-sensitivity solar conversion panel that converts room light into electrical energy for powering the calculator. The widespread use of these devices has increased industrial productivity considerably since their introduction in the 1970s.

1.4.2 Modern Calculator Techniques

Finding Natural Trigonometric Functions. The natural trigonometric functions of *all* angles are obtained easily, with great speed and precision.

EXAMPLE. Find the natural trigonometric function of sin 26°41′26″.

First, convert from degrees, minutes, and seconds to decimal degrees (see Sec. 1.3.1):

$$26°41'26'' = 26.690555°$$

1.30 CHAPTER ONE

Press: sin
Enter: 26.690555, then =
Answer: 0.4491717 (the natural function)

The natural sine, cosine, and tangent of any angle may thus be found. Negative angles are found by pressing sin, cos, or tan; entering the decimal degrees; changing sign to minus; and then pressing =.

The cotangent, secant, and cosecant are found by using the reciprocal button (x^{-1}) on the calculator.

Finding Common and Natural Logarithms of Numbers. The common, or Briggs, logarithm system is constructed with a base of 10 (see Sec. 1.2.1).

EXAMPLE

$$10^1 = 10 \quad \text{and} \quad \log_{10} 10 = 1$$

$$10^2 = 100 \quad \text{and} \quad \log_{10} 100 = 2$$

$$10^3 = 1000 \quad \text{and} \quad \log_{10} 1000 = 3$$

Therefore, $\log_{10} 110.235$ is found by pressing log and entering the number into the calculator:

Press: log
Enter: 110.235, then =
Answer: 2.042319506

Since the logarithmic value is the exponent to which 10 is raised to obtain the number, we will perform this calculation:

$$10^{2.042319506} = 110.235$$

PROOF

Enter: 10
Press: y^x
Enter: 2.042319506
Press: =
Answer: 110.2349999, or 110.235 to three decimal places.

The natural, or hyperbolic, logarithm of a number is found in a similar manner.

EXAMPLE. Find the natural, or hyperbolic, logarithm of 110.235.

Press: ln
Enter: 110.235, then =
Answer: 4.702614451

MATHEMATICS FOR MACHINISTS AND METALWORKERS   1.31

Powers and Roots (Exponentials). Finding powers and roots (exponentials) of numbers is simple on the pocket calculator and renders logarithmic procedures and tables of logarithms obsolete, as well as the functions of numbers tables found in outdated handbooks.

EXAMPLE. Find the square root of 3.4575.

Press: $\sqrt{x}$
Enter: 3.4575, then =
Answer: 1.859435398

The procedure takes but a few seconds.

EXAMPLE. Find $(0.0625)^4$.

Enter: 0.0625
Press: x^y
Enter: 4
Press: =
Answer: 1.525879×10^{-5}

EXAMPLE. Find the cube root of 5.2795, or $(5.2795)^{1/3}$.

Enter: 3   or   *Enter:* 5.2795
Press: $x\sqrt{y}$ *Press:* x^y
Enter: 5.2795   *Enter:* 0.33333
Press: =   *Press:* =
Answer: 1.7412626   *Answer:* 1.74126

NOTE. Radicals written in exponential notation:

$$\sqrt[3]{5} = (5)^{1/3} = (5)^{0.33333}$$

$$\sqrt{6} = (6)^{1/2} = (6)^{0.5}$$

$$\sqrt[3]{(6.245)^2} = (6.245)^{2/3} = (6.245)^{0.66666}$$

1.4.3 Pocket Calculator Bracketing Procedures

When entering an equation into the pocket calculator, correct bracketing procedures must be used in order to prevent calculation errors. An incorrect procedure results in a SYN ERROR or MATH ERROR message on the calculator display, or an incorrect numerical answer.

EXAMPLES. x = unknown to be calculated.

Equation	Enter as Shown, Then Press = or EXE
$x = \dfrac{a+b}{c-d}$	$(a+b)/(c-d) =$ or EXE
$x = \dfrac{(6 \times 7)+1}{\pi + 7}$	$((6 \times 7)+1)/(\pi+7)$
$x = \dfrac{(a+b)/2}{(a \times d)+2}$	$(a+b)/2/((a \times d)+2)$
$x = \dfrac{(1/\tan 40)+2}{1/(2 \times \sin 30)}$	$(1/\tan 40)+2/1/(2 \sin 30)$
$x = \dfrac{2.215 \times 4.188 \times 6.235}{2+d}$	$(2.215)(4.188)(6.235)/(2+d)$
$x = \dfrac{b}{c-d}$	$b/(c-d)$

The examples shown are some of the more common types of bracketing. The bracketing will become more difficult on long, complex equations. Explanations of the order of entry and the bracketing procedures are usually shown in the instruction book that comes with the pocket calculator. A calculator that displays the equation as it is being entered into the calculator is the preferred type. The Casio calculator shown in Fig. 1.31 is of this type. The more advanced TI and HP calculators shown in Fig. 1.32 also display the entire entered equation, making them easier to use and reducing the chance of bracket entry error.

1.5 ANGLE CONVERSIONS— DEGREES AND RADIANS

Converting Degrees to Radians and Radians to Degrees. To convert from degrees to radians, you must first find the degrees as decimal degrees (see previous section). If R represents radians, then

$$2\pi R = 360° \quad \text{or} \quad \pi R = 180°$$

From this,

$$1 \text{ radian} = \frac{180}{\pi} = 57.2957795°$$

And

$$1° = \frac{\pi}{180} = 0.0174533 \text{ radian}$$

EXAMPLE. Convert 56.785° to radians.

$$56.785 \times 0.0174533 = 0.9911 \text{ radian}$$

So

$$56.785° = 0.9911 \text{ radian}$$

EXAMPLE. Convert 2.0978R to decimal degrees.

$$57.2957795 \times 2.0978 = 120.0591°$$

So

$$2.0978 \text{ radians} = 120.0591°$$

See the radians and degrees template—Fig. 1.33.

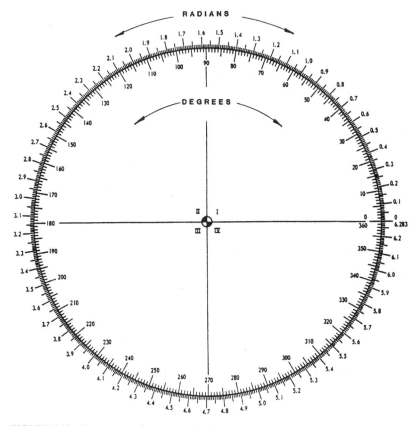

FIGURE 1.33 Degrees to radians conversion chart.

Important Mathematical Constants

$\pi = 3.1415926535898$

1 radian = $57.295779513082°$

$1° = 0.0174532925199$ radian

$2\pi R = 360°$

$\pi R = 180°$

1 radian = $180/\pi°$

$1° = \pi/180$ radians

$e = 2.718281828$ (base of natural logarithms)

1.6 POWERS-OF-TEN NOTATION

Numbers written in the form 1.875×10^5 or 3.452×10^{-6} are so stated in powers-of-ten notation. Arithmetic operations on numbers which are either very large or very small are easily and conveniently processed using the powers-of-ten notation and procedures. This process is automatically carried out by the hand-held scientific calculator. If the calculated answer is larger or smaller than the digital display can handle, the answer will be given in powers-of-ten notation.

This method of handling numbers is always used in scientific and engineering calculations when the values of the numbers so dictate. Engineering notation is usually given in multiples of 3, such as 1.246×10^3, 6.983×10^{-6}, etc.

How to Calculate with Powers-of-Ten Notation. Numbers with many digits may be expressed more conveniently in powers-of-ten notation, as shown here.

$$0.000001389 = 1.389 \times 10^{-6}$$

$$3{,}768{,}145 = 3.768145 \times 10^6$$

You are actually counting the number of places that the decimal point is shifted, either to the right or to the left. Shifting to the right produces a negative exponent, and shifting to the left produces a positive exponent.

Multiplication, division, exponents, and radicals in powers-of-ten notation are easily handled, as shown here.

$$1.246 \times 10^4 \, (2.573 \times 10^{-4}) = 3.206 \times 10^0 = 3.206 \quad \text{(Note: } 10^0 = 1\text{)}$$

$$1.785 \times 10^7 \div (1.039 \times 10^{-4}) = (1.785/1.039) \times 10^{7-(-4)} = 1.718 \times 10^{11}$$

$$(1.447 \times 10^5)^2 = (1.447)^2 \times 10^{10} = 2.094 \times 10^{10}$$

$$\sqrt{1.391 \times 10^8} = 1.391^{1/2} \times 10^{8/2} = 1.179 \times 10^4$$

In the preceding examples, you must use the standard algebraic rules for addition, subtraction, multiplication, and division of exponents or powers of numbers.

MATHEMATICS FOR MACHINISTS AND METALWORKERS   1.35

Thus,

- Exponents are algebraically added for multiplication.
- Exponents are algebraically subtracted for division.
- Exponents are algebraically multiplied for power raising.
- Exponents are algebraically divided for taking roots.

1.7 PERCENTAGE CALCULATIONS

Percentage calculation procedures have many applications in machining, design, and metalworking problems. Although the procedures are relatively simple, it is easy to make mistakes in the manipulations of the numbers involved.

Ordinarily, 100 percent of any quantity is represented by the number 1.00, meaning the total quantity. Thus, if we take 50 percent of any quantity, or any multiple of 100 percent, it *must* be expressed as a decimal:

$$1\% = 0.01$$

$$10\% = 0.10$$

$$65.5\% = 0.655$$

$$145\% = 1.45$$

In effect, we are dividing the percentage figure, such as 65.5 percent, by 100 to arrive at the decimal equivalent required for calculations.

Let us take a percentage of a given number:

$$45\% \text{ of } 136.5 = 0.45 \times 136.5 = 61.425$$

$$33.5\% \text{ of } 235.7 = 0.335 \times 235.7 = 78.9595$$

Let us now compare two arbitrary numbers, 33 and 52, as an illustration:

$$\frac{52-33}{33} = 0.5758$$

Thus, the number 52 is 57.58 percent larger than the number 33. We also can say that 33 increased by 57.58 percent is equal to 52; that is, $0.5758 \times 33 + 33 = 52$. Now,

$$\frac{52-33}{52} = 0.3654$$

Thus, the number 52 minus 36.54 percent of itself is 33. We also can say that 33 is 36.54 percent less than 52, that is, $0.3654 \times 52 = 19$ and $52 - 19 = 33$. The number 33 is what percent of 52? That is, $33/52 = 0.6346$. Therefore, 33 is 63.46 percent of 52.

Example of a Practical Percentage Calculation. A spring is compressed to 417 lbf and later decompressed to 400 lbf, or load. The percentage pressure drop is (417 − 400)/417 = 0.0408, or 4.08 percent. The pressure, or load, is then increased to 515 lbf. The percentage increase over 400 lbf is therefore (515 − 400)/515 = 0.2875, or 28.75 percent.

Percentage problem errors are quite common, even though the calculations are simple. In most cases, if you remember that the divisor is the number of which you want the percentage, either increasing or decreasing, the simple errors can be avoided. Always *back-check* your answers using the percentages against the numbers.

1.8 TEMPERATURE SYSTEMS AND CONVERSIONS

There are four common temperature systems used in engineering and design calculations: Fahrenheit (°F), Celsius (formerly centigrade; °C), Kelvin (K), and Rankine (°R). The conversion equation for Celsius to Fahrenheit or Fahrenheit to Celsius is

$$\frac{5}{9} = \frac{°C}{°F - 32}$$

This exact relational equation is all that you need to convert from either system. Enter the known temperature, and solve the equation for the unknown value.

EXAMPLE. You wish to convert 66°C to Fahrenheit.

$$\frac{5}{9} = \frac{66}{°F - 32}$$

$$5°F - 160 = 594$$

$$°F = 150.8$$

This method is much easier than trying to remember the two equivalent equations, which are:

$$°C = \frac{5}{9}(°F - 32)$$

and

$$°F = \frac{9}{5}°C + 32$$

The other two systems, Kelvin and Rankine, are converted as described here. The Kelvin and Celsius scales are related by

$$K = 273.18 + °C$$

Thus, 0°C = 273.18 K. Absolute zero is equal to −273.18°C.

EXAMPLE. A temperature of $-75°C = 273.18 + (-75°C) = 198.18$ K.
The Rankine and Fahrenheit scales are related by

$$°R = 459.69 + °F$$

Thus, $0°F = 459.69°R$. Absolute zero is equal to $-459.69°F$.

EXAMPLE. A temperature of $75°F = 459.69 + (75°F) = 534.69°R$.

1.9 DECIMAL AND MILLIMETER EQUIVALENTS

See Fig. 1.34.

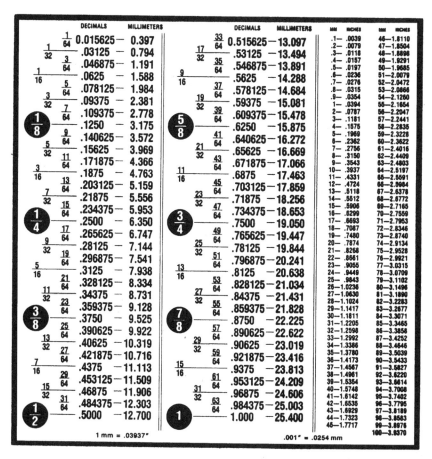

FIGURE 1.34 Decimal and millimeter equivalents.

1.10 SMALL WEIGHT EQUIVALENTS: U.S. CUSTOMARY (GRAINS AND OUNCES) VERSUS METRIC (GRAMS)

1 gram	= 15.43 grains
1 gram	= 15,430 milligrains
1 pound	= 7000 grains
1 ounce	= 437.5 grains
1 ounce	= 28.35 grams
1 grain	= 0.0648 grams
1 grain	= 64.8 milligrams
0.1 grain	= 6.48 milligrams
1 micrograin	= 0.0000648 milligrams
1000 micrograins	= 0.0648 milligrams
1 grain	= 0.002286 ounces
10 grains	= 0.02286 ounces or 0.648 grams
100 grains	= 0.2286 ounces or 6.48 grams

EXAMPLE. To obtain the weight in grams, multiply the weight in grains by 0.0648. Or, divide the weight in grains by 15.43.

EXAMPLE. To obtain the weight in grains, multiply the weight in grams by 15.43. Or, divide the weight in grams by 0.0648.

1.11 MATHEMATICAL SIGNS AND SYMBOLS

TABLE 1.1 Mathematical Signs and Symbols

$+$	Plus, positive
$-$	Minus, negative
$\times$ or $\cdot$	Times, multiplied by
$\div$ or $/$	Divided by
$=$	Is equal to
$\equiv$	Is identical to
$\cong$	Is congruent to or approximately equal to
$\sim$	Is approximately equal to or is similar to
$<$ and $\not<$	Is less than, is not less than
$>$ and $\not>$	Is greater than, is not greater than
$\neq$	Is not equal to
$\pm$	Plus or minus, respectively
$\mp$	Minus or plus, respectively
$\propto$	Is proportional to
$\rightarrow$	Approaches, e.g., as $x \rightarrow 0$
$\leq, \leqq$	Less than or equal to
$\geq, \geqq$	More than or equal to
$\therefore$	Therefore
$:$	Is to, is proportional to
Q.E.D.	Which was to be proved, end of proof
%	Percent
#	Number
@	At
$\angle$ or $\measuredangle$	Angle
° ′ ″	Degrees, minutes, seconds
$\|$, $//$	Parallel to
$\perp$	Perpendicular to
e	Base of natural logs, 2.71828 . . .
π	Pi, 3.14159 . . .
( )	Parentheses
[ ]	Brackets
{ }	Braces
′	Prime, $f'(x)$
″	Double prime, $f''(x)$
$\sqrt{\ }, \sqrt[n]{\ }$	Square root, nth root
$1/x$ or x^{-1}	Reciprocal of x
!	Factorial
∞	Infinity
Δ	Delta, increment of
∂	Curly d, partial differentiation
Σ	Sigma, summation of terms
Π	The product of terms, product
arc	As in arcsine (the angle whose sine is)
f	Function, as $f(x)$
rms	Root mean square
$\|x\|$	Absolute value of x
i	For -1
j	Operator, equal to -1

TABLE 1.2 The Greek Alphabet

α	A	alpha	ι	I	iota	ρ	P	rho
β	B	beta	κ	K	kappa	σ	Σ	sigma
γ	Γ	gamma	λ	Λ	lambda	τ	T	tau
δ	Δ	delta	μ	M	mu	υ	Y	upsilon
ε	E	epsilon	ν	N	nu	φ	Φ	phi
ζ	Z	zeta	φ	Ξ	xi	χ	X	chi
η	H	eta	o	O	omicron	ψ	Ψ	psi
θ	Θ	theta	π	Π	pi	ω	Ω	omega

CHAPTER 2
MENSURATION OF PLANE AND SOLID FIGURES

2.1 MENSURATION

Mensuration is the mathematical name for calculating the areas, volumes, length of sides, and other geometric parts of standard geometric shapes such as circles, spheres, polygons, prisms, cylinders, cones, etc., through the use of mathematical equations or formulas. Included here are the most frequently used and important mensuration formulas for the common geometric figures, both plane and solid. (See Figs. 2.1 through 2.36.)

Symbols
A	area
a, b, etc.	sides
A, B, C	angles
h	height perpendicular to base b
L	length of side or edge
r	radius
n	number of sides
C	circumference
V	volume
S	surface area

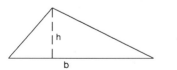

$$A = \frac{1}{2}bh$$

FIGURE 2.1 Oblique triangle.

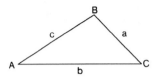

$$A = \frac{1}{2}ab \sin C$$

$$A = \sqrt{s(s-a)(s-b)(s-c)}$$

where $s = \frac{1}{2}(a+b+c)$

FIGURE 2.2 Oblique triangle.

$$A = ab$$

FIGURE 2.3 Rectangle.

$$A = bh$$

FIGURE 2.4 Parallelogram.

$$A = \frac{1}{2}cd$$

FIGURE 2.5 Rhombus.

$$A = \frac{1}{2}(a+b)h$$

FIGURE 2.6 Trapezoid.

MENSURATION OF PLANE AND SOLID FIGURES 2.3

Surfaces and Volumes of Polyhedra:
(Where L = leg or edge)

Polyhedron	Surface	Volume
Tetrahedron	$1.73205L^2$	$0.11785L^3$
Hexahedron	$6L^2$	$1L^3$
Octahedron	$3.46410L^2$	$0.47140L^3$

FIGURE 2.7 Polyhedra.

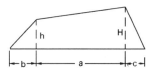

$$A = \frac{(H+h)a + bh + cH}{2}$$

FIGURE 2.8 Trapezium.

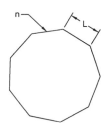

In a polygon of n sides of L, the radius of the inscribed circle is:

$$r = \frac{L}{2} \cot \frac{180}{n} ;$$

The radius of the circumscribed circle is:

$$r_1 = \frac{L}{2} \csc \frac{180}{n}$$

FIGURE 2.9 Regular polygon.

The radius of a circle inscribed in any triangle whose sides are a, b, c is:

$$r = \frac{\sqrt{s(s-a)(s-b)(s-c)}}{s}$$

where $s = \frac{1}{2}(a+b+c)$

FIGURE 2.10 Inscribed circle.

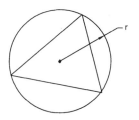

In any triangle, the radius of the circumscribed circle is:

$$r = \frac{abc}{4\sqrt{s\,(s-a)(s-b)(s-c)}}$$

where $s = \frac{1}{2}(a+b+c)$

FIGURE 2.11 Circumscribed circle.

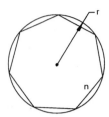

FIGURE 2.12 Inscribed polygon.

Area of an inscribed polygon is:

$$A = \frac{1}{2} nr^2 \sin \frac{2\pi}{n}$$

where r = radius of circumscribed circle
       n = number of sides

FIGURE 2.13 Circumscribed polygon.

Area of a circumscribed polygon is:

$$A = nr^2 \tan \frac{\pi}{n}$$

where r = radius of inscribed circle
       n = number of sides

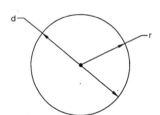

FIGURE 2.14 Circle—circumference.

$$C = 2\pi r = \pi d$$

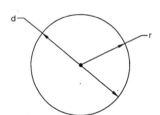

FIGURE 2.15 Circle—area.

$$A = \pi r^2 = \frac{1}{4}\pi d^2$$

MENSURATION OF PLANE AND SOLID FIGURES 2.5

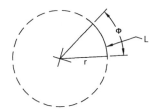

Length of arc L:

$$L = \frac{\pi r \phi}{180} \quad \text{(when } \phi \text{ is in degrees)}$$

$$L = \pi \phi \quad \text{(when } \phi \text{ is in radians)}$$

FIGURE 2.16 Length of arc.

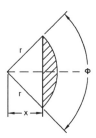

Length of chord:

$$AB = 2r \sin \frac{1}{2}\phi$$

Area of the sector:

$$A = \frac{\pi r^2 \phi}{360} = \frac{rL}{2}$$

where L = length of the arc

FIGURE 2.17 Chord and sector.

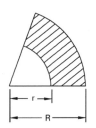

Area of segment of a circle:

$$A = \frac{\pi r^2 \phi}{360} - \frac{r^2 \sin\phi}{2}$$

where: $\phi = 180° - 2 \arcsin\left(\dfrac{x}{r}\right)$

If ϕ is in radians:

$$A = \frac{1}{2} r^2 (\phi - \sin\phi)$$

FIGURE 2.18 Segment of a circle.

Area of the ring between circles. Circles need not be concentric:

$$A = \pi(R + r)(R - r)$$

FIGURE 2.19 Ring.

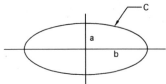

FIGURE 2.20 Ellipse.

Circumference and area of an ellipse (approximate):

$$C = 2\pi \sqrt{\frac{a^2 + b^2}{2}}$$

Area:
$$A = \pi ab$$

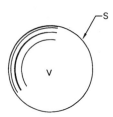

FIGURE 2.21 Pyramid.

Volume of a pyramid:

$$V = \frac{1}{3} \times \text{area of base} \times h$$

where h = altitude

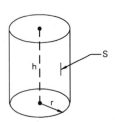

FIGURE 2.22 Sphere.

Surface and volume of a sphere:
$$S = 4\pi r^2 = \pi d^2$$
$$V = \frac{4}{3}\pi r^3 = \frac{1}{6}\pi d^3$$

FIGURE 2.23 Cylinder.

Surface and volume of a cylinder:
$$S = 2\pi rh$$
$$V = \pi r^2 h$$

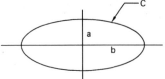

FIGURE 2.24 Cone.

Surface and volume of a cone:
$$S = \pi r \sqrt{r^2 + h^2}$$
$$V = \frac{\pi}{3} r^2 h$$

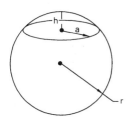

FIGURE 2.25 Spherical segment.

Area and volume of a curved surface of a spherical segment:

$$A = 2\pi rh \quad V = \left(\frac{\pi h^2}{3}\right)(3r - h)$$

When a is radius of base of segment:

$$V = \frac{\pi h}{4}(h^2 + 3a^2)$$

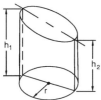

FIGURE 2.26 Frustum of a cone.

Surface area and volume of a frustum of a cone:

$$S = \pi (r_1 + r_2) \sqrt{h^2 + (r_1 - r_2)^2}$$

$$V = \frac{h}{3}(r_1^2 + r_1 r_2 + r_2^2)\pi$$

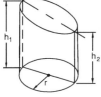

FIGURE 2.27 Truncated cylinder.

Area and volume of a truncated cylinder:

$$A = \pi r (h_1 + h_2)$$

$$V = \frac{\pi}{2} r^2 (h_1 + h_2)$$

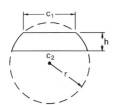

FIGURE 2.28 Spherical zone.

Area and volume of a spherical zone:
$A = 2\pi rh$

$$V = \frac{\pi}{6} h \left(\frac{3c_1^2}{4} + \frac{3c_2^2}{4} + h^2\right)$$

FIGURE 2.29 Spherical wedge.

Area and volume of a spherical wedge:

$$A = \frac{\phi}{360} 4\pi r^2$$

$$V = \frac{\phi}{360} \cdot \frac{4\pi r^3}{3}$$

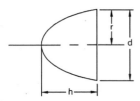

FIGURE 2.30 Paraboloid.

Volume of a paraboloid:

$$V = \frac{\pi r^2 h}{2}$$

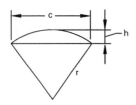

FIGURE 2.31 Spherical sector.

Area and volume of a spherical sector (yields total area):

$$A = \pi r \left(2h + \frac{c}{2}\right)$$

$$V = \frac{2\pi r^2 h}{3} \quad c = 2\sqrt{h(2r-h)}$$

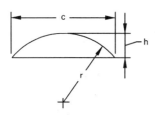

FIGURE 2.32 Spherical segment.

Area and volume of a spherical segment:

$$A = 2\pi rh$$

$$\text{Spherical surface} = \pi\left(\frac{c^2}{4} + h^2\right)$$

$$c = 2\sqrt{h(2r-h)} \quad r = \frac{c^2 + 4h^2}{8h}$$

$$V = \pi h^2 \left(r - \frac{h}{3}\right)$$

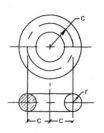

FIGURE 2.33 Torus.

Area and volume of a torus:
$A = 4\pi^2 cr$ (total surface)
$V = 2\pi^2 cr^2$ (total volume)

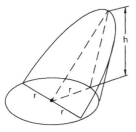

Area and volume of a portion of a cylinder (base edge = diameter):

$$A = 2rh \quad V = \frac{2}{3} r^2 h$$

FIGURE 2.34 Portion of a cylinder.

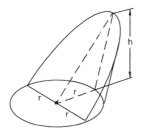

Area and volume of a portion of a cylinder (special cases):

$$A = \frac{h(ad \pm c \times \text{perimeter of base})}{r \pm c}$$

$$V = \frac{h\left(\dfrac{2}{3} a^3 \pm cA\right)}{r \pm c}$$

where d = diameter of base circle

FIGURE 2.35 Special case of a cylinder.

Note. Use $+c$ when base area is larger than half the base circle; use $-c$ when base area is smaller than half the base circle.

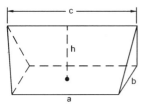

Volume of a wedge

$$V = \frac{(2b + c)ah}{6}$$

FIGURE 2.36 Wedge.

2.2 PROPERTIES OF THE CIRCLE

See Fig. 2.37.

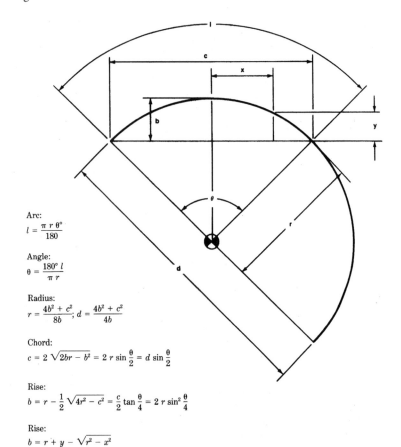

Arc:
$$l = \frac{\pi r \theta°}{180}$$

Angle:
$$\theta = \frac{180° \, l}{\pi r}$$

Radius:
$$r = \frac{4b^2 + c^2}{8b}; \, d = \frac{4b^2 + c^2}{4b}$$

Chord:
$$c = 2\sqrt{2br - b^2} = 2r \sin \frac{\theta}{2} = d \sin \frac{\theta}{2}$$

Rise:
$$b = r - \frac{1}{2}\sqrt{4r^2 - c^2} = \frac{c}{2} \tan \frac{\theta}{4} = 2r \sin^2 \frac{\theta}{4}$$

Rise:
$$b = r + y - \sqrt{r^2 - x^2}$$
where $y = b - r + \sqrt{r^2 - x^2}$ and $x = \sqrt{r^2 - (r + y - b)^2}$.

FIGURE 2.37 Properties of the circle

CHAPTER 3
LAYOUT PROCEDURES FOR GEOMETRIC FIGURES

3.1 GEOMETRIC CONSTRUCTION

The following figures show the methods used to perform most of the basic geometric constructions used in standard drawing and layout practices. Many of these constructions have widespread use in the machine shop, the sheet metal shop, and in engineering.

- *To divide any straight line into any number of equal spaces* (Fig. 3.1). To divide line *AB* into five equal spaces, draw line *AC* at any convenient angle such as angle *BAC*. With a divider or compass, mark off five equal spaces along line *AC* with a divider or compass. Now connect point 5 on line *AC* with the endpoint of line *AB*. Draw line *CB*, and parallel transfer the other points along line *AC* to intersect line *AB*, thus dividing it into five equal spaces.

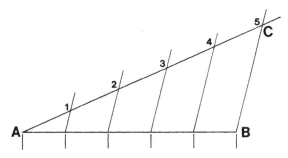

FIGURE 3.1 Dividing a line equally.

- *To bisect any angle BAC* (Fig. 3.2), swing an arc from point *A* through points *d* and *e*. Swing an arc from point *d* and another equal arc from point *e*. The intersection of these two arcs will be at point *f*. Draw a line from point *A* to point *f*, forming the bisector line *AD*.

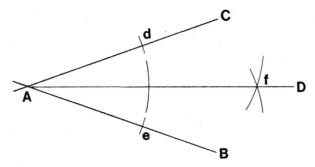

FIGURE 3.2 Bisecting an angle.

- *To divide any line into two equal parts and erect a perpendicular* (Fig. 3.3), draw an arc from point A that is more than half the length of line AB. Using the same arc length, draw another arc from point B. The intersection points of the two arcs meet at points c and d. Draw the perpendicular bisector line cd.

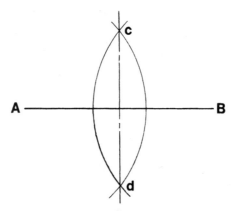

FIGURE 3.3 Erecting a perpendicular.

- *To erect a perpendicular line through any point along a line* (Fig. 3.4), from point c along line AB, mark points 1 and 2 equidistant from point c. Select an arc length on the compass greater than the distance from points 1 to c or points 2 to c. Swing this arc from point 1 and point 2. The intersection of the arcs is at point f. Draw a line from point f to point c, which is perpendicular to line AB.

LAYOUT PROCEDURES FOR GEOMETRIC FIGURES 3.3

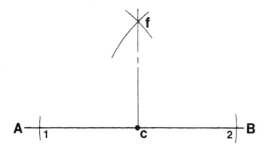

FIGURE 3.4   Perpendicular to a point.

- *To draw a perpendicular to a line AB, from a point f, a distance from it* (Fig. 3.5), with point f as a center, draw a circular arc intersecting line AB at points c and d. With points c and d as centers, draw circular arcs with radii longer than half the distance between points c and d. These arcs intersect at point e, and line fe is the required perpendicular.

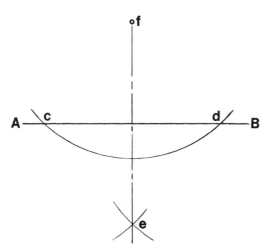

FIGURE 3.5   Drawing a perpendicular to a line from a point.

- *To draw a circular arc with a given radius through two given points* (Fig. 3.6), with points A and B as centers and the set given radius, draw circular arcs intersecting at point f. With point f as a center, draw the circular arc which will intersect both points A and B.

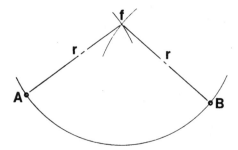

FIGURE 3.6 Drawing a circular arc through given points.

- *To find the center of a circle or the arc of a circle* (Fig. 3.7), select three points on the perimeter of the given circle such as A, B, and C. With each of these points as a center and the same radius, describe arcs which intersect each other. Through the points of intersection, draw lines *fb* and *fd*. The intersection point of these two lines is the center of the circle or circular arc.

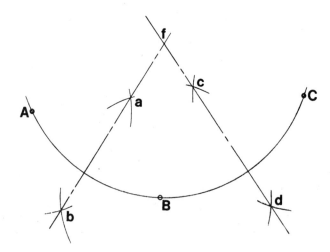

FIGURE 3.7 Finding the center of a circle.

- *To draw a tangent to a circle from any given point on the circumference* (Fig. 3.8), through the tangent point *f*, draw a radial line *OA*. At point *f*, draw a line *CD* at right angles to *OA*. Line *CD* is the required tangent to point *f* on the circle.
- *To draw a geometrically correct pentagon within a circle* (Fig. 3.9), draw a diameter *AB* and a radius *OC* perpendicular to it. Bisect *OB* and with this point *d* as center and a radius *dC*, draw arc *Ce*. With center *C* and radius *Ce*, draw arc *ef*. *Cf* is then a side of the pentagon. Step off distance *Cf* around the circle using a divider.

LAYOUT PROCEDURES FOR GEOMETRIC FIGURES

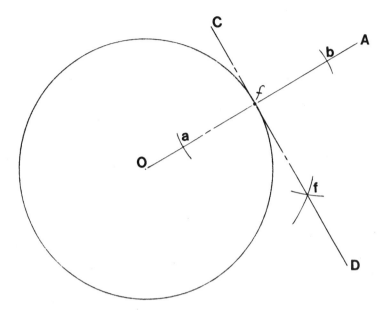

FIGURE 3.8 Drawing a tangent to a given point on a circle.

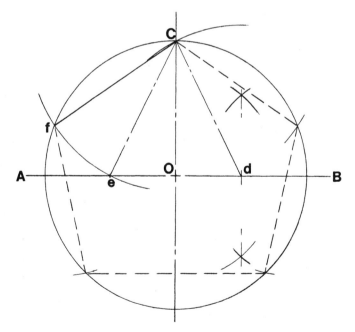

FIGURE 3.9 Drawing a pentagon.

- *To draw a geometrically correct hexagon given the distance across the points* (Fig. 3.10), draw a circle on *ab* with *a* diameter. With the same radius, *Of,* and with points 6 and 3 as centers, draw arcs intersecting the circle at points 1, 2, 4, and 5, and connect the points.

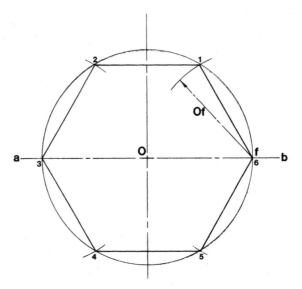

FIGURE 3.10 Drawing a hexagon.

- *To draw a geometrically correct octagon in a square* (Fig. 3.11), draw the diagonals of the square. With the corners of the square *b* and *d* as centers and a radius of half the diagonal distance *Od,* draw arcs intersecting the sides of the square at points 1 through 8, and connect these points.

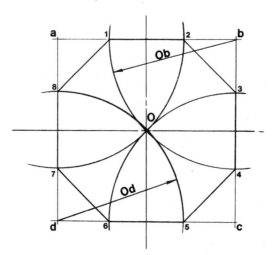

FIGURE 3.11 Drawing an octagon.

LAYOUT PROCEDURES FOR GEOMETRIC FIGURES   3.7

- *Angles of the pentagon, hexagon, and octagon* (Fig. 3.12).

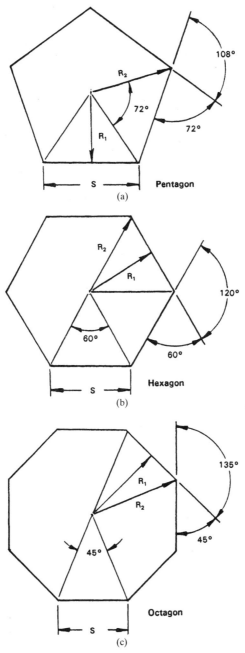

FIGURE 3.12 (*a*) Angles of the pentagon. (*b*) Hexagon. (*c*) Octagon.

- *To draw an ellipse given the major and minor axes* (Fig. 3.13). The concentric-circle method: On the two principle diameters *ef* and *cd* which intersect at point *O*, draw circles. From a number of points on the outer circle, such as *g* and *h*, draw radii *Og* and *Oh* intersecting the inner circle at points *g'* and *h'*. From *g* and *h*, draw lines parallel to *Oa*, and from *g'* and *h'*, draw lines parallel to *Od*. The intersection of the lines through *g* and *g'* and *h* and *h'* describe points on the ellipse. Each quadrant of the concentric circles may be divided into as many equal angles as required or as dictated by the size and accuracy required.

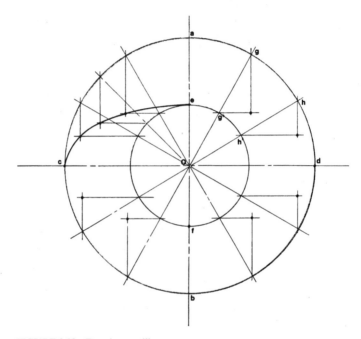

FIGURE 3.13 Drawing an ellipse.

- *To draw an ellipse using the parallelogram method* (Fig. 3.14), on the axes *ab* and *cd*, construct a parallelogram. Divide *aO* into any number of equal parts, and divide *ae* into the same number of equal parts. Draw lines through points 1 through 4 from points *c* and *d*. The intersection of these lines will be points on the ellipse.
- *To draw a parabola using the parallelogram method* (Fig. 3.15), divide *Oa* and *ba* into the same number of equal parts. From the divisions on *ab*, draw lines converging at *O*. Lines drawn parallel to line *OA* and intersecting the divisions on *Oa* will intersect the lines drawn from point *O*. These intersections are points on the parabola.

LAYOUT PROCEDURES FOR GEOMETRIC FIGURES 3.9

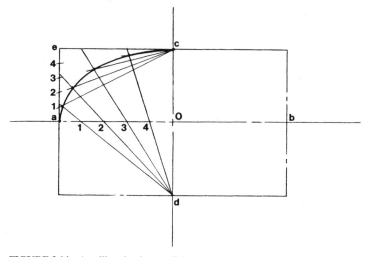

FIGURE 3.14 An ellipse by the parallelogram method.

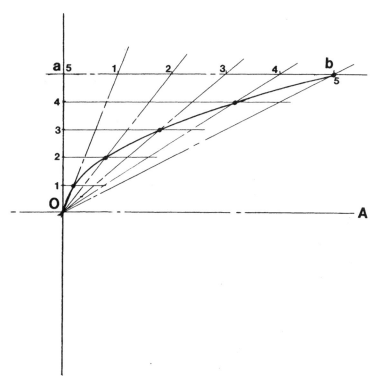

FIGURE 3.15 A parabola by the parallelogram method.

- *To draw a parabola using the offset method* (Fig. 3.16), the parabola may be plotted by computing the offsets from line $O5$. These offsets vary as the square of their distance from point O. If $O5$ is divided into five equal parts, distance $1e$ will be $\frac{1}{25}$ distance $5a$. Offset $2d$ will be $\frac{4}{25}$ distance $5a$; offset $3c$ will be $\frac{9}{25}$ distance $5a$, etc.

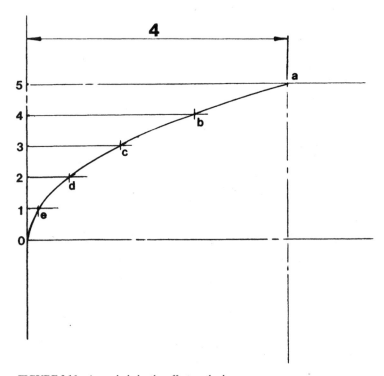

FIGURE 3.16 A parabola by the offset method.

- *To draw a parabolic envelope* (Fig. 3.17), divide Oa and Ob into the same number of equal parts. Number the divisions from Oa and Ob, 1 through 6, etc. The intersection of points 1 and 6, 2 and 5, 3 and 4, 4 and 3, 5 and 2, and 6 and 1 will be points on the parabola. This parabola's axis is not parallel to either ordinate.
- *To draw a parabola when the focus and directrix are given* (Fig. 3.18), draw axis Op through point f and perpendicular to directrix AB. Through any point k on the axis Op, draw lines parallel to AB. With distance kO as a radius and f as a center, draw an arc intersecting the line through k, thus locating a point on the parabola. Repeat for Oj, Oi, etc.

LAYOUT PROCEDURES FOR GEOMETRIC FIGURES 3.11

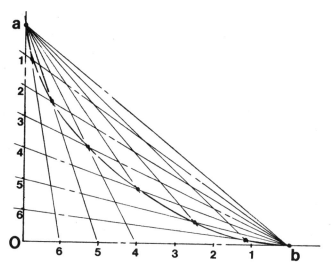

FIGURE 3.17 A parabolic envelope.

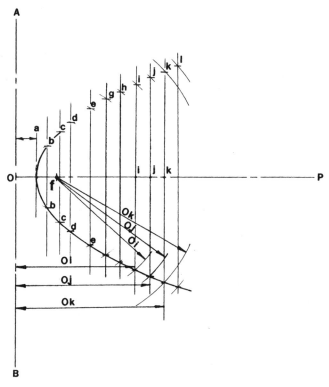

FIGURE 3.18 A parabolic curve.

- *To draw a helix* (Fig. 3.19), draw the two views of the cylinder and measure the lead along one of the contour elements. Divide the lead into a number of equal parts, say 12. Divide the circle of the front view into the same number of equal parts, say 12. Project points 1 through 12 from the top view to the stretch-out of the helix in the right view. Angle ϕ is the *helix angle,* whose tangent is equal to $L/\pi D$, where L is the lead and D is the diameter.
- *To draw the involute of a circle* (Fig. 3.20), divide the circle into a convenient number of parts, preferably equal. Draw tangents at these points. Line $a2$ is perpendicular to radial line $O2$, line $b3$ is perpendicular to radial line $O3$, etc. Lay off on these tangent lines the true lengths of the arcs from the point of tangency to the starting point, 1. For accuracy, the true lengths of the arcs may be calculated (see Fig. 2.37 in the chapter on mensuration for calculating arc lengths). The involute of the circle is the basis for the involute system of gearing. Another method for finding points mathematically on the involute is shown in Sec. 7.1.
- *To draw the spiral of Archimedes* (Fig. 3.21), divide the circle into a number of equal parts, drawing the radii and assigning numbers to them. Divide the radius $O8$ into the same number of equal parts, numbering from the center of the circle. With O as a center, draw a series of concentric circles from the marked points on

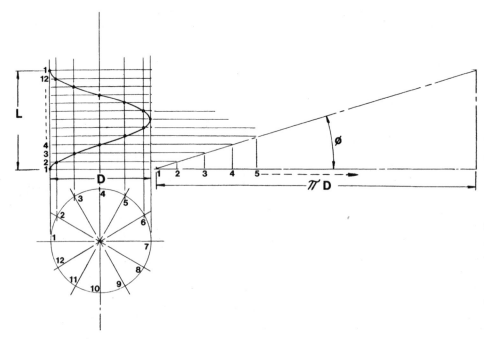

FIGURE 3.19   To draw a helix.

LAYOUT PROCEDURES FOR GEOMETRIC FIGURES 3.13

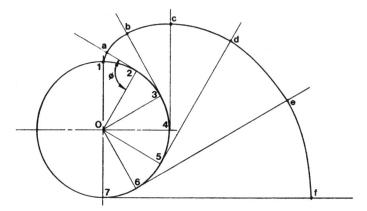

FIGURE 3.20 To draw the involute of the circle.

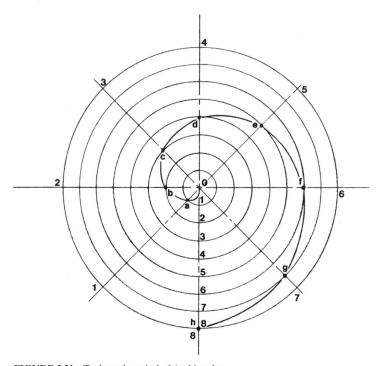

FIGURE 3.21 To draw the spiral of Archimedes.

the radius, 1 through 8. The spiral curve is defined by the points of intersection of the radii and the concentric circles at points *a, b, c, d, e, f, g,* and *h*. Connect the points with a smooth curve. The Archimedean spiral is the curve of the heart cam, which is used to convert uniform rotary motion into uniform reciprocating motion. See Chap. 8 on ratchets and cam geometry.

CHAPTER 4
MEASUREMENT AND CALCULATION PROCEDURES FOR MACHINISTS

4.1 SINE BAR AND SINE PLATE CALCULATIONS

Sine Bar Procedures. Referring to Figs. 4.1a and b, find the sine bar setting height for an angle of 34°25′ using a 5-in sine bar.

$$\sin 34°25' = \frac{x}{5} \quad (34°25' = 34.416667 \text{ decimal degrees})$$

$$\sin 34.416667° = \frac{x}{5}$$

$$x = 5 \times 0.565207$$

$$x = 2.826 \text{ in}$$

Set the sine bar height with Jo-blocks or precision blocks to 2.826 in.

From this example it is apparent that the setting height can be found for any sine bar length simply by multiplying the length of the sine bar times the natural sine value of the required angle. The simplicity, speed, and accuracy possible for setting sine bars with the aid of the pocket calculator renders sine bar tables obsolete. No sine bar table will give you the required setting height for such an angle as 42°17′26″, but by using the calculator procedure, this becomes a routine, simple process with less chance for error.

Method

1. Convert the required angle to decimal degrees.
2. Find the natural sine of the required angle.
3. Multiply the natural sine of the angle by the length of the sine bar to find the bar setting height (see Fig. 4.1b).

4.2 CHAPTER FOUR

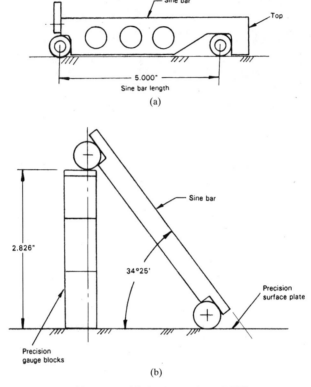

FIGURE 4.1 (*a*) Sine bar. (*b*) Sine bar setting at 34°25′.

Formulas for Finding Angles. Refer to Fig. 4.2 when angles α and φ are known to find angles *X, A, B,* and *C.*

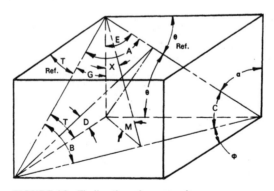

FIGURE 4.2 Finding the unknown angles.

NOTE. $\angle (\theta) = 90° - \angle \alpha$

$$\angle G = 90° - \angle T$$

$$\angle A + \angle B + \angle C = 180°$$

$$\angle X + \angle M = 90°$$

$$\tan X = \tan \alpha \cos \phi$$

$$\sin C = \frac{\cos \alpha}{\cos X}$$

Angle $B = 180° - $ (angle A + angle C)

$$\tan A = \frac{\sin \alpha \sin C}{\sin \phi - (\sin \alpha \cos C)}$$

D = true angle

$$\tan D = \tan \phi \sin \Theta$$

$$\tan C = \frac{\sin D}{\tan \Theta}$$

$$\tan M = \sqrt{(\tan \Theta)^2 + (\tan T)^2}$$

$$\cos A = \cos E \cos G$$

$$\cos A = \sin \Theta \sin T$$

Formulas and Development for Finding True and Apparent Angles. See Fig. 4.3*a*, where α = apparent angle, Θ = true angle, and ϕ = angle of rotation.

NOTE. Apparent angle α is *OA* triangle projected onto plane *OB*. See also Fig. 4.3*b*.

$$\tan \Theta = \frac{K}{L}$$

$$\tan \alpha = \frac{K}{L \cos \phi}$$

$$\tan \alpha \cos \phi = \frac{K}{L}$$

$$\frac{K}{L} = \tan \Theta = \cos \phi \tan \alpha$$

4.4 CHAPTER FOUR

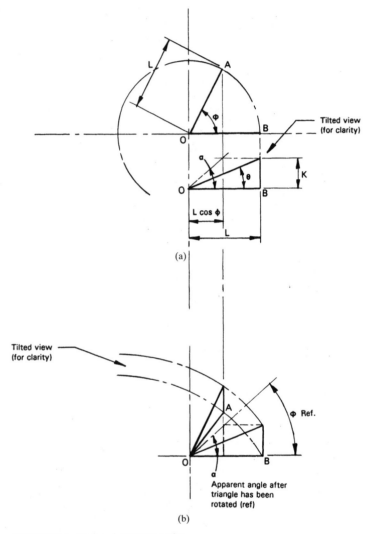

FIGURE 4.3 True and apparent angles.

or
$$\tan \Theta = \cos \phi \tan \alpha$$

and
$$\tan \alpha = \frac{\tan \Theta}{\cos \phi}$$

MEASUREMENT AND CALCULATION PROCEDURES FOR MACHINISTS   4.5

The three-dimensional relationships shown for the angles and triangles in the preceding figures and formulas are of importance and should be understood. This will help in the setting of compound sine plates when it is required to set a compound angle.

Setting Compound Sine Plates. For setting two known angles at 90° to each other, proceed as shown in Figs. 4.4a, b, and c.

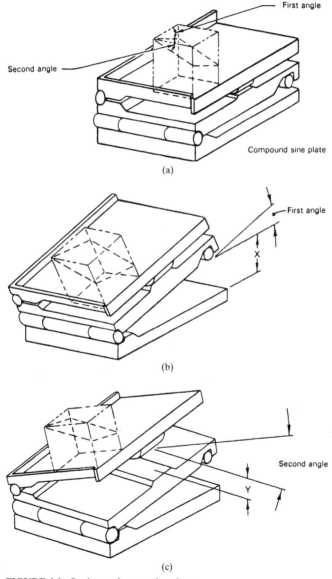

FIGURE 4.4 Setting angles on a sine plate.

EXAMPLE. First angle = 22.45°. Second angle = 38.58° (see Fig. 4.4). To find the amount the intermediate plate must be raised from the base plate (X dimension in Fig. 4.4b) to obtain the desired first angle,

1. Find the natural cosine of the second angle (38.58°), and multiply this times the natural tangent of the first angle (22.45°).
2. Find the arctangent of this product, and then find the natural sine of this angle.
3. This natural sine is now multiplied by the length of the sine plate to find the X dimension in Fig. 4.4b to which the intermediate plate must be set.
4. Set up the Jo-blocks to equal the X dimension, and set in position between base plate and intermediate plate.

EXAMPLE

$$\cos 38.58° = 0.781738$$

$$\tan 22.45° = 0.413192$$

$$0.781738 \times 0.413192 = 0.323008$$

$$\arctan 0.323008 = 17.900872°$$

$$\sin 17.900872° = 0.307371$$

$$0.307371 \times 10 \text{ in (for 10-in sine plate)} = 3.0737 \text{ in}$$

Therefore, set X dimension to 3.074 in (to three decimal places).

To find the amount the top plate must be raised (Y dimension in Fig. 4.4c) above the intermediate plate to obtain the desired second angle,

1. Find the natural sine of the second angle, and multiply this times the length of the sine plate.
2. Set up the Jo-blocks to equal the Y dimension, and set in position between the top plate and the intermediate plate.

EXAMPLE

$$\sin 38.58° = 0.632607$$

$$0.632607 \times 10 \text{ in (for 10-in sine plate)} = 6.32607$$

Therefore, set Y dimension to 6.326 in (to three decimal places).

4.2 SOLUTIONS TO PROBLEMS IN MACHINING AND METALWORKING

The following sample problems will show in detail the importance of trigonometry and basic algebraic operations as apply to machining and metalworking. By using the

MEASUREMENT AND CALCULATION PROCEDURES FOR MACHINISTS 4.7

methods and procedures shown in Chap. 1 and this chapter of the handbook, you will be able to solve many basic and complex machining and metalworking problems.

Taper (Fig. 4.5). Solve for x if y is given; solve for y if x is given; solve for d. Use the tangent function:

$$\tan A = \frac{y}{x}$$

$$d = D - 2y$$

where A = taper angle
D = outside diameter of rod
d = diameter at end of taper
x = length of taper
y = drop of taper

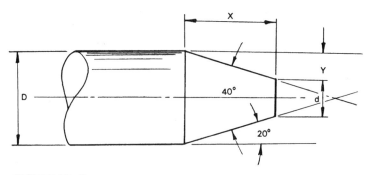

FIGURE 4.5 Taper.

EXAMPLE. If the rod diameter = 0.9375 diameter, taper length = 0.875 = x, and taper angle = 20° = angle A, find y and d from

$$\tan 20° = \frac{y}{x}$$

$$y = x \tan 20°$$

$$= 0.875(0.36397)$$

$$= 0.318$$

$$d = D - 2y$$

$$= 0.9375 - 2(0.318)$$

$$= 0.9375 - 0.636$$

$$= 0.3015$$

Countersink Depths (Three Methods for Calculating). See Fig. 4.6.

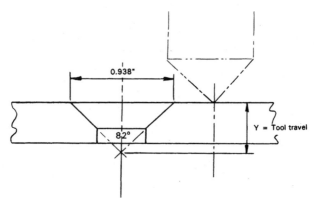

FIGURE 4.6 Countersink depth.

Method 1. To find the tool travel y from the top surface of the part for a given countersink finished diameter at the part surface,

$$y = \frac{D/2}{\tan \tfrac{1}{2}A} \quad \text{(Fig. 4.6)}$$

where D = finished countersink diameter
  A = countersink angle
  y = tool advance from surface of part

$$y = \frac{0.938/2}{\tan 41°} = \frac{0.469}{0.869} = 0.5397, \text{ or } 0.540$$

Method 2. To find the tool travel from the edge of the hole (Fig. 4.7) where D = finished countersink diameter, H = hole diameter, and A = ½ countersink angle, 41°,

$$\tan A = \frac{x}{y}$$

$$y = \frac{x}{\tan A} \quad \text{or} \quad \frac{x}{\tfrac{1}{2} \text{ countersink angle}}$$

First, find x from

$$D = H + 2x$$

If D = 0.875 and H = 0.500,

$$0.875 = 0.500 + 2x$$

$$2x = 0.375$$

$$x = 0.1875$$

MEASUREMENT AND CALCULATION PROCEDURES FOR MACHINISTS   4.9

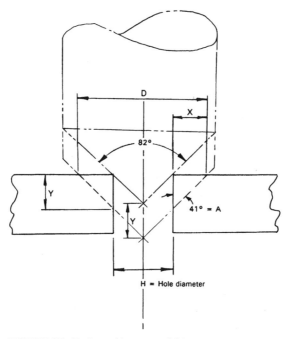

FIGURE 4.7   Tool travel in countersinking.

Now, solve for y, the tool advance:

$$y = \frac{x}{\tan A}$$

$$= \frac{0.1875}{\tan 41°}$$

$$= \frac{0.1875}{0.8693}$$

$$= 0.2157, \text{ or } 0.216 \text{ (tool advance from edge of hole)}$$

Method 3.   To find tool travel from edge of hole (Fig. 4.8) where D = finished countersink diameter, d = hole diameter, ϕ = ½ countersink angle, and H = countersink tool advance from edge of hole,

$$H = \tfrac{1}{2}(D - d) \cotan \phi \quad \text{or} \quad H = \frac{D - d}{2 \tan \phi}$$

(Remember that $\cotan \phi = 1/\tan \phi$ or $\tan \phi = 1/\cotan \phi$.)

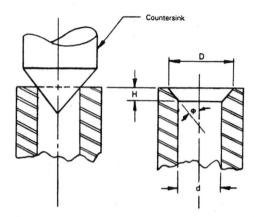

FIGURE 4.8 Tool travel from the edge of the hole, countersinking.

Finding Taper Angle α. Given dimensions shown in Fig. 4.9, find angle α and length x.

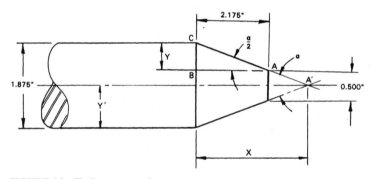

FIGURE 4.9 Finding taper angle α.

First, find angle α from

$$y = \frac{1.875 - 0.500}{2} = \frac{1.375}{2} = 0.6875$$

Then solve triangle ABC for ½ angle α:

$$\tan \frac{1}{2}\alpha = \frac{0.6875}{2.175} = 0.316092$$

$$\arctan \tfrac{1}{2}\alpha = 0.316092$$

$$\tfrac{1}{2}\alpha = 17.541326°$$

$$\alpha = 2 \times 17.541326°$$

$$\text{angle } \alpha = 35.082652°$$

Then solve triangle $A'B'C$, where $y' = 0.9375$ or ½ diameter of rod:

$$\text{Angle } C = 90° - 17.541326°$$
$$= 72.458674°$$

Now the x dimension is found from

$$\tan \frac{1}{2}\alpha = \frac{0.9375}{x}$$

$$x = \frac{0.9375}{\tan \tfrac{1}{2} \alpha}$$

$$= \frac{0.9375}{0.316092}$$

$$= 2.966 \text{ (side } A'B' \text{ or length } x)$$

Geometry of the Pentagon, Hexagon, and Octagon. The following figures show in detail how basic trigonometry and algebra are used to formulate the solutions to these geometric figures.
The Pentagon. See Fig. 4.10.

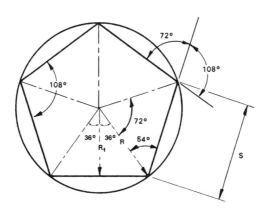

FIGURE 4.10 Pentagon geometry.

Where R = radius of circumscribed circle
 R_1 = radius of inscribed circle
 S = length of side

From the law of sines, we know the following relation:

$$\frac{S}{\sin 72°} = \frac{R}{\sin 54°}$$

$$S \sin 54° = R \sin 72°$$

$$S = \frac{R \sin 72}{\sin 54}$$

$$= \frac{R(0.9511)}{0.8090}$$

$$= 1.1756R \text{ (where } R = \text{radius of circumscribed circle)}$$

Also,

$$S = \frac{R_1 \sin 72°}{\cos 36 \sin 54} \quad \left(\textbf{Note: } \cos 36° = \frac{R_1}{R}\right)$$

$$= \frac{R_1(0.9511)}{(0.8090)(0.8090)}$$

$$= \frac{R_1(0.9511)}{0.6545}$$

$$= 1.4532 R_1 \text{ (where } R_1 = \text{radius of inscribed circle)}$$

The area of the pentagon is thus

$$A_1 = \frac{1}{2}\left(\frac{S}{2}\right)R_1$$

$$= \frac{SR_1}{4}$$

$$= \frac{S(R \cos 36)}{4} \quad \left(\textbf{Note: } \cos 36 = \frac{R_1}{R} \text{ and } R_1 = R \cos 36\right)$$

$$A_T = 5\left(\frac{SR_1}{4}\right)$$

$$= 1.25 SR_1 \text{ (the total area of the pentagon)}$$

The Hexagon. See Fig. 4.11.

Where R = radius of inscribed circle
 R_1 = radius of circumscribed circle
 S = length of side
 W = width across points

From Fig. 4.11 we know the following relation:

MEASUREMENT AND CALCULATION PROCEDURES FOR MACHINISTS   **4.13**

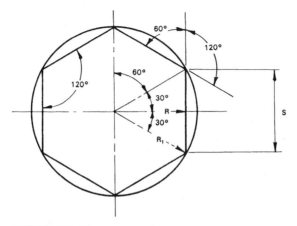

FIGURE 4.11 Hexagon geometry.

$$\tan 30° = \frac{x}{R} \quad \text{and} \quad S = 2x \quad \text{or} \quad x = \frac{S}{2}$$

$$x = R \tan 30$$

Then $S = 2R \tan 30$

$\qquad = 2R(0.57735)$

$\qquad = 1.1457R$

$$\cos 30° = \frac{R}{R_1}$$

$$R = R_1 \cos 30$$

$$R_1 = \frac{R}{\cos 30}$$

$$R_1 = \frac{R}{0.86605} \quad \text{or} \quad R_1 = 1.15467R$$

and $W = 2(1.15467)R$

$\qquad = 2.30934R$ (diameter of the circumscribed circle)

Area: $A = 2.598S^2$

$\qquad = 3.464r^2$

$\qquad = 2.598R_1^2$

The Octagon. See Fig. 4.12.

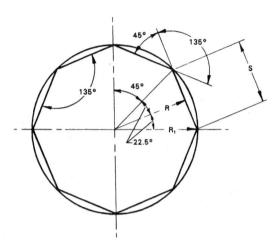

FIGURE 4.12 Octagon geometry.

Where R = radius of inscribed circle
 R_1 = radius of circumscribed circle
 S = length of side
 W = width across points

From Fig. 4.12 we know the following relation:

$$\tfrac{1}{2}S = R \tan 22°30'$$

$$S = 2R \tan 22°30'$$

$$S = 2R(0.414214)$$

$$S = 0.828428R$$

Also: $R = 1.20711S$

Then, $\cos 22°30' = \dfrac{R}{R_1}$

$$R = R_1 \cos 22°30'$$

$$R_1 = \dfrac{R}{\cos 22°30'}$$

Then, $W = 2\left(\dfrac{R}{\cos 22°30'}\right)$

$$= 2.165R$$

Area: $A = 4.828S^2$

$= 3.314r^2$

In the preceding three figures of the pentagon, hexagon, and octagon, you may calculate the other relationships between S, R, and R_1 as required using the procedures shown as a guide. When one of these parts is known, the other parts may be found in relation to the given part.

4.3 CALCULATIONS FOR SPECIFIC MACHINING PROBLEMS (TOOL ADVANCE, TAPERS, NOTCHES AND PLUGS, DIAMETERS, RADII, AND DOVETAILS)

Drill-Point Advance. When drilling a hole, it is often useful to know the distance from the cylindrical end of the drilled hole to the point of the drill for any angle point and any diameter drill. Refer to Fig. 4.13, where the advance t is calculated from

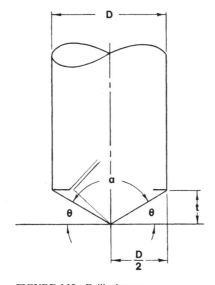

FIGURE 4.13 Drill advance.

$$\tan\left(\frac{180 - \alpha}{2}\right) = \frac{t}{D/2}$$

Then

$$t = \frac{D}{2} \tan\left(\frac{180 - \alpha}{2}\right)$$

where D = diameter of drill, in
 α = drill-point angle

EXAMPLE. What is the advance t for a 0.875-in-diameter drill with a 118° point angle?

$$t = \frac{0.875}{2} \tan\left(\frac{180 - 118}{2}\right) \quad \text{Note:} \quad \frac{180 - \alpha}{2} = \angle \Theta \text{ (reference)}$$

$$= 0.4375 \tan 31°$$

$$= 0.4375 \times 0.60086 = 0.2629 \text{ in}$$

Tapers. Finding taper angles under a variety of given conditions is an essential part of machining mathematics. Following are a variety of taper problems with their associated equations and solutions.

For taper in inches per foot, see Fig. 4.14a. If the taper in inches per foot is denoted by T, then

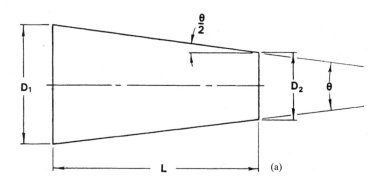

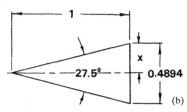

FIGURE 4.14 Taper angles.

$$T = \frac{12(D_1 - D_2)}{L}$$

where D_1 = diameter of larger end, in
 D_2 = diameter of smaller end, in
 L = length of tapered part along axis, in
 T = taper, in/ft

Also, to find the angle Θ, use the relationship

$$\tan \Theta = \frac{12(D_1 - D_2)}{L}$$

then find arctan Θ for angle Θ.

EXAMPLE. $D_1 = 1.255$ in, $D_2 = 0.875$ in, and $L = 3.5$ in. Find angle Θ.

$$\tan \Theta = \frac{1.255 - 0.875}{3.5} = \frac{0.380}{0.875} = 0.43429$$

$$= 0.43429$$

And arctan $0.43429 = 23.475°$ or $23°28.5'$.

Figure 4.14b shows a taper angle of 27.5° in 1 in, and the taper per inch is therefore 0.4894. This is found simply by solving the triangle formed by the axis line, which is 1 in long, and half the taper angle, which is 13.75°. Solve one of the right-angled triangles formed by the tangent function:

$$\tan 13.75° = \frac{x}{1}$$

and $\quad x = \tan 13.75° = 0.2447$

and $\quad 2 \times 0.2447 = 0.4894$

as shown in Fig. 4.14b.

The taper in inches per foot is equal to 12 times the taper in inches per inch. Thus, in Fig. 4.14b, the taper per foot is $12 \times 0.4894 = 5.8728$ in.

Typical Taper Problems

1. Set two disks of known diameter and a required taper angle at the correct center distance L (see Fig. 4.15).

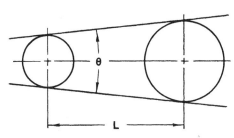

FIGURE 4.15 Taper.

Given: Two disks of known diameter d and D and the required angle Θ. Solve for L.

$$L = \frac{D-d}{2\left(\sin \dfrac{\Theta}{2}\right)}$$

2. Find the angle of the taper when given the taper per foot (see Fig. 4.16).

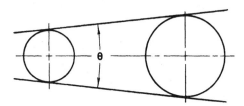

FIGURE 4.16 Angle of taper.

Given: Taper per foot T. Solve for angle Θ.

$$\Theta = 2\left(\arctan \frac{T}{24}\right)$$

3. Find the taper per foot when the diameters of the disks and the length between them are known (see Fig. 4.17).

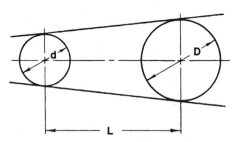

FIGURE 4.17 Taper per foot.

Given: d, D, and L. Solve for T.

$$T = \tan\left(\arcsin \frac{D-d}{L}\right) \times 24$$

4. Find the angle of the taper when the disk dimensions and their center distance is known (see Fig. 4.18).

MEASUREMENT AND CALCULATION PROCEDURES FOR MACHINISTS   4.19

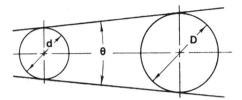

FIGURE 4.18   Angle of taper.

Given: $d, D,$ and $L.$ Solve for angle Θ.

$$\Theta = 2\left(\arcsin \frac{D-d}{2L}\right)$$

5. Find the taper in inches per foot measured at right angles to one side when the disk diameters and their center distance are known (see Fig. 4.19).

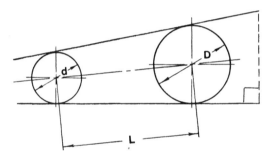

FIGURE 4.19   Taper in inches per foot.

Given: $d, D,$ and $L.$ Solve for $T,$ in inches per foot.

$$T = \tan\left[2\left(\arcsin \frac{D-d}{2L}\right)\right] \times 12$$

6. Set a given angle with two disks in contact when the diameter of the smaller disk is known (see Fig. 4.20).

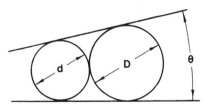

FIGURE 4.20   Setting a given angle.

Given: d and Θ. Solve for D, diameter of the larger disk.

$$D = \left(\frac{2d \sin \dfrac{\Theta}{2}}{1 - \sin \dfrac{\Theta}{2}}\right) + d$$

Figure 4.21 shows an angle-setting template which may be easily constructed in any machine shop. Angles of extreme precision are possible to set using this type of tool. The diameters of the disks may be machined precisely, and the center distances between the disks may be set with a gauge or Jo-blocks. Also, any angle may be repeated when a record is kept of the disk diameters and the precise center distance. The angle Θ, taper per inch, or taper per foot may be calculated using some of the preceding equations.

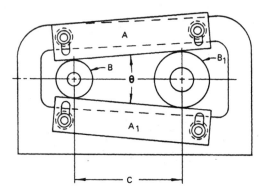

FIGURE 4.21 Angle-setting template.

Checking Angles and Notches with Plugs. A machined plug may be used to check the correct width of an angular opening or machined notch or to check templates or parts which have corners cut off or in which the body is notched with a right angle. This is done using the following techniques and simple equations.

In Figs. 4.22, 4.23, and 4.24, $D = a + b - c$ (right-angle notches). To check the width of a notched opening, see Fig. 4.25 and the following equation:

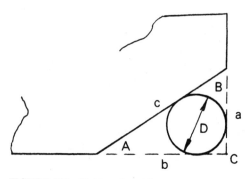

FIGURE 4.22 Right-angle notch.

MEASUREMENT AND CALCULATION PROCEDURES FOR MACHINISTS 4.21

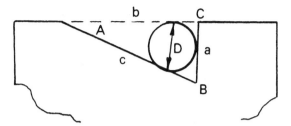

FIGURE 4.23 Right-angle notch.

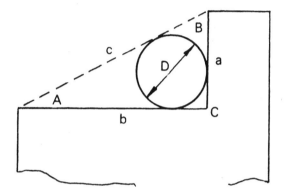

FIGURE 4.24 Right-angle notch.

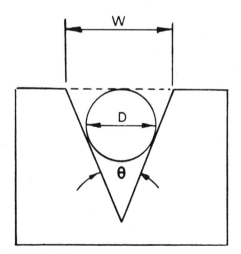

FIGURE 4.25 Width of notched opening.

$$D = W \tan\left(45° - \frac{\Theta}{2}\right)$$

When the correct size plug is inserted into the notch, it should be tangent to the opening indicated by the dashed line.

Also, the equation for finding the correct plug diameter that will contact all sides of an oblique or non-right-angle triangular notch is as follows (see Fig. 4.26):

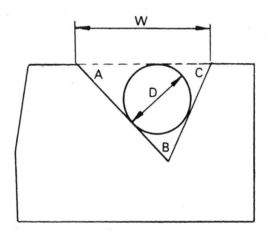

FIGURE 4.26 Finding plug diameter.

$$D = \frac{2W}{\left(\cot\frac{A}{2}\right) + \left(\cot\frac{C}{2}\right)} \quad \text{or} \quad 2W\left(\tan\frac{A}{2} + \tan\frac{C}{2}\right)$$

where W = width of notch, in
  A = angle A
  B = angle B

Finding Diameters. When the diameter of a part is too large to measure accurately with a micrometer or vernier caliper, you may use a 90° or any convenient included angle on the tool (which determines angle A) and measure the height H as shown in Fig. 4.27. The simple equation for calculating the diameter D for any angle A is as follows:

$$D = H\frac{2}{\csc A - 1} \quad \text{(\textbf{Note:} } \csc 45° = 1.4142\text{)}$$

Thus, the equation for measuring the diameter D with a 90° square reduces to

$$D = 4.828H$$

Then, if the height H measured was 2.655 in, the diameter of the part would be

$$D = 4.828 \times 2.655 = 12.818 \text{ in}$$

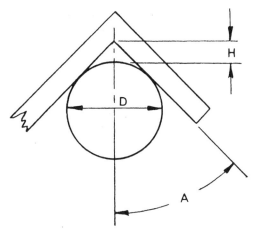

FIGURE 4.27 Finding the diameter.

When measuring large gears, a more convenient angle for the measuring tool would be 60°, as shown in Fig. 4.28. In this case, the calculation becomes simple. When the measuring angle of the tool is 60° (angle $A = 30°$), the diameter D of the part is $2H$.

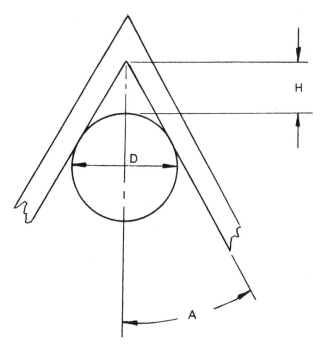

FIGURE 4.28 Finding the diameter.

For measuring either inside or outside radii on any type of part, such as a casting or a broken segment of a wheel, the calculation for the radius of the part is as follows (see Figs. 4.29 and 4.30):

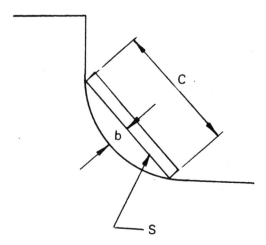

FIGURE 4.29 Finding the radius.

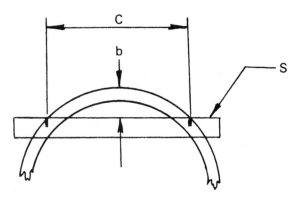

FIGURE 4.30 Finding the radius.

$$r = \frac{4b^2 + c^2}{8b}$$

where r = radius of part, in
 b = chordal height, in
 c = chord length, in
 S = straight edge

The chord should be made from a precisely measured piece of tool steel flat, and the chordal height b may be measured with an inside telescoping gauge or micrometer.

Measuring Radius of Arc by Measuring over Rolls or Plugs. Another accurate method of finding or checking the radius on a part is illustrated in Figs. 4.31 and 4.32. In this method, we may calculate either an inside or an outside radius by the following equations:

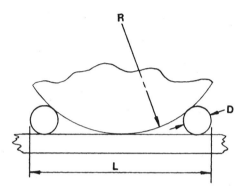

FIGURE 4.31 Finding the radius.

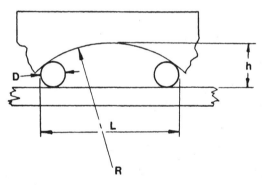

FIGURE 4.32 Finding the radius.

$$r = \frac{(L+D)^2}{8D} \quad \text{(for convex radii, Fig. 4.31)}$$

$$r = \frac{(L+D)^2}{8(h-D)} + \frac{h}{2} \quad \text{(for concave radii, Fig. 4.32)}$$

where L = length over rolls or plugs, in
 D = diameter of rolls or plugs, in
 h = height of concave high point above the rolls or plugs, in

For accuracy, the rolls or plugs must be placed on a tool plate or plane table and the distance L across the rolls measured accurately. The diameter D of the rolls or plugs also must be measured precisely and the height h measured with a telescoping gauge or inside micrometers.

Measuring Dovetail Slides. The accuracy of machining of dovetail slides and their given widths may be checked using cylindrical rolls (such as a drill rod) or wires and the following equations (see Figs. 4.33a and b):

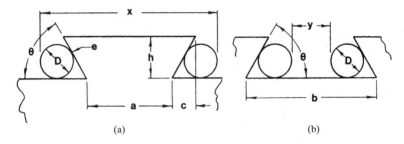

FIGURE 4.33 Measuring dovetail slides.

$$x = D\left(\cot \frac{\Theta}{2}\right) + a \quad \text{(for male dovetails, Fig. 4.33a)}$$

$$y = b - D\left(1 + \cot \frac{\Theta}{2}\right) \quad \text{(for female dovetails, Fig. 4.33b)}$$

NOTE. $c = h \cot \Theta$. Also, the diameter of the rolls or wire should be sized so that the point of contact e is below the corner or edge of the dovetail.

Taper Problem and Calculation Procedures. Figure 4.34 shows a typical machined part with two intersecting tapers. The given or known dimensions are shown here, and it is required to solve for the unknown dimensions and the weight of the part in ounces, after machining.

Given: L_1, R_2, d_1, angle α, and angle β.

Find: R_1, R_3, bc, d_2, L_2, L_3, and L_4; then calculate the volume and weight of the part, when the material is specified.

$L_1 = 6.000$ in, $R_2 = 0.250$ in, $d_1 = 0.875$ in, angle $\alpha = 15°$, and angle $\beta = 60°$.
Solution.

$R_2 \times 2 = d_2 \qquad\qquad R_3 = \dfrac{d_1}{2} = \dfrac{0.875}{2} = 0.4375$ in

$0.250 \times 2 = d_2 = 0.500$ in

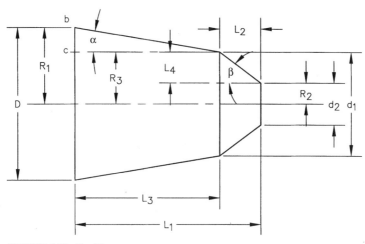

FIGURE 4.34 Double taper.

$$L_4 = \frac{d_1}{2} - 0.250 \qquad \tan \alpha = \frac{bc}{L_3}$$

$$L_4 = \frac{0.875}{2} - 0.250 \qquad bc = L_3 \tan \alpha$$

$L_4 = 0.4375 - 0.250 \qquad bc = 5.892 \times \tan 15°$

$L_4 = 0.1875 \text{ in} \qquad bc = 5.892 \times 0.2680$

$\qquad\qquad\qquad\qquad bc = 1.579 \text{ in}$

$$\tan \beta = \frac{L_4}{L_2} \qquad R_1 = R_3 + bc$$

$$L_2 = \frac{L_4}{\tan \beta} \qquad R_1 = 0.4375 + 1.579$$

$$L_2 = \frac{0.1875}{\tan 60°} = \frac{0.1875}{1.732} \qquad R_1 = 2.017 \text{ in}$$

$L_2 = 0.108 \text{ in}$

$L_3 = L_1 - L_2 \qquad D = 2R_1$

$L_3 = 6.000 - 0.108 \qquad D = 2 \times 2.017$

$L_3 = 5.892 \text{ in} \qquad D = 4.034 \text{ in dia.}$

From Fig. 4.35, the volume and weight of the machined tapered part can be calculated as follows.

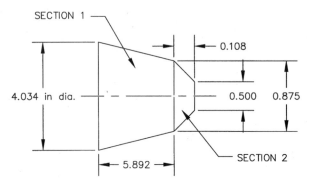

FIGURE 4.35 Volume of double taper part.

Per the dimensions given in Fig. 4.35, find the volume in cubic inches and the part weight, when the part is made from 7075-T651 aluminum alloy stock:
Solution.   The part consists of two sections, both of which are frustums of a cone. The equation for calculating the volume of a frustum of a cone is:

$$V = \frac{h}{3}\left(r_1^2 + r_1 r_2 + r_2^2\right)\pi$$

Section 1: $r_1 = 2.017, r_2 = 0.438,$ and $h = L_3 = 5.892$

$$V_1 = \frac{5.892}{3}(2.017^2 + 2.017 \times 0.438 + 0.438^2)3.1416$$

$V_1 = 1.964(4.068 + 0.883 + 0.192)3.1416$

$V_1 = 1.964 \times 5.143 \times 3.1416$

$V_1 = 31.733$ in^3

Section 2: $r_1 = 0.438, r_2 = 0.250,$ and $h = 0.108$

$$V_2 = \frac{0.108}{3}(0.438^2 + 0.438 \times 0.250 + 0.250^2)3.1416$$

$V_2 = 0.036(0.192 + 0.110 + 0.063)3.1416$

$V_2 = 0.036 \times 0.365 \times 3.1416$

$V_2 = 0.041$ in^3

Volume of the part = $V_1 + V_2$

$V_{\text{total}} = 31.733 + 0.041$

$V_{\text{total}} = 31.774$ in^3

Since 7075-T651 aluminum alloy weighs 0.101 lb/in³, the part weighs:

W = volume, in³ × density of 7075-T651

$W = 31.774 \times 0.101$

$W = 3.21$ lb or 51.35 oz

Find the diameter of a tapered end for a given radius r (see Fig. 4.36).

Problem. To find the diameter d, when the radius r and angle of taper α are known:

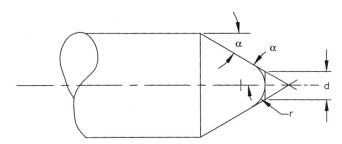

FIGURE 4.36 Finding diameter d.

Given: $\alpha = 25°, r = 0.250$ in

Using the equation:

$$d = 2r\left(\cot \frac{90° + \alpha}{2}\right)$$

solve for d:

$$d = 2 \times 0.250\left(\cot \frac{90° + 25°}{2}\right)$$

$$d = 0.500\left(\cot \frac{115°}{2}\right)$$

$$d = 0.500(\cot 57.5°)$$

$$d = 0.500\left(\frac{1}{\tan 57.5°}\right)$$

$$d = 0.500\left(\frac{1}{1.570}\right)$$

$$d = 0.500 \times 0.637$$

$$d = 0.319 \text{ in}$$

Checking the Angle of a Tapered Part by Measuring over Cylindrical Pins

Problem. Calculate what the measurement L over pins should be, when the diameter of the pins is 0.250 in, and the angle α on the machined part is given as 41° (see Fig. 4.37).

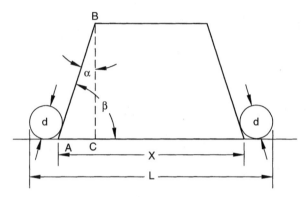

FIGURE 4.37 Checking the angle of a tapered part.

Solution. With an X dimension of 2.125 in, $d = 0.250$ in, and angle $\alpha = 41°$, the solution for the measured distance L can be found by using the following equation:

$$L = X + d\left[\tan\left(\frac{90° - \alpha}{2}\right) + 1\right] \quad \text{angle } \beta = 90° - \alpha$$

$$L = 2.125 + 0.250\left[\tan\left(\frac{90° - 41°}{2}\right) + 1\right]$$

$$L = 2.125 + 0.250[\tan(24.5°) + 1]$$

$$L = 2.125 + 0.250(1.456)$$

$$L = 2.125 + 0.364$$

$$L = 2.489 \text{ in}$$

If the L dimension is measured as 2.502 in, and X remains 2.125 in, calculate for the new angle α_1 using the transposed equation:

$$\alpha_1 = 90° + 2 \arctan\left(\frac{-L + X + d}{d}\right) \quad \text{(see MathCad in Sec. 1.2.2)}$$

$$\alpha_1 = 90° + 2 \arctan\left(\frac{-2.502 + 2.125 + 0.250}{0.250}\right)$$

$$\alpha_1 = 90° + 2 \arctan\left(\frac{-0.127}{0.250}\right)$$

MEASUREMENT AND CALCULATION PROCEDURES FOR MACHINISTS **4.31**

$\alpha_1 = 90° + 2 \arctan(-0.508)$

$\alpha_1 = 90° + 2(-26.931°)$

$\alpha_1 = 90° - 53.862°$

$\alpha_1 = 36.138°$ or $36°08'16.8''$

NOTE. The MathCad-generated equations are the transpositions of the basic equation, set up to solve for α, X, and d. These equations were calculated symbolically for these other variables in the basic equation; Sec. 1.2.2 shows the results both when the angles are given in degrees and when the angles are given in radians.

Note. There are 2π rad in $360°$, 1 rad $= (180/\pi)°$; $1° = (\pi/180)$ rad. That is, 1 rad $= 57.2957795°$; $1° = 0.0174533$ rad.

Forces and Vector Forces on Taper Keys or Wedges. Refer to Figs. 4.38a and b and the following equations to determine the forces on taper keys and wedges.
For Fig. 4.38a we have:

δ = angle of friction = $\arctan \mu$; $\tan \delta = \mu$, or $\tan^{-1} \mu = \delta$

μ = coefficient of friction (you must know or estimate this coefficient prior to solving the equations, because δ depends on μ, the coefficient of friction at the taper key or wedge surfaces). The coefficient of friction of steel on steel is generally taken as 0.150 to 0.200.

$$\text{Efficiency}, \eta = \frac{\tan \alpha}{\tan(\alpha + 2\delta)}$$

$$P = \frac{F\eta}{\tan \alpha}$$

$$F = \frac{P \tan \alpha}{\eta}$$

$$N = \frac{F\eta}{\sin \alpha}$$

For Fig. 4.38b we have:

$$P = \frac{F\eta}{2 \tan \alpha}$$

$$F = \frac{2P \tan \alpha}{\eta}$$

$$N = \frac{F\eta}{2 \sin \alpha}$$

4.32 CHAPTER FOUR

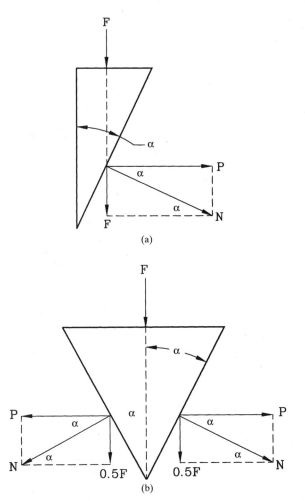

FIGURE 4.38 (*a*) Forces of a single tapered wedge; (*b*) forces of a double tapered wedge.

In Fig. 4.39, for milling cutter angles $\alpha = 20°$, $\beta = 45°$, cutter nose radius of 0.125 in, and a groove width $x = 0.875$ in, we can solve for the plunge depth y, and the distance d to the tool vertical centerline, using the following equations:

$$y = \frac{x \cos \alpha \cos \beta}{\sin (\alpha + \beta)} - r\left\{\left[\frac{1}{\sin\left(\frac{\alpha + \beta}{2}\right)}\right] \cos \left(\frac{\beta - \alpha}{2}\right) - 1\right\} \quad \text{(Eq. 4.1)}$$

Bracket the equation in the calculator as follows:

($x \cos \alpha \cos \beta/\sin (\alpha + \beta)$) − r((1/sin (($\alpha + \beta$)/2)) (cos (($\beta - \alpha$)/2)) − 1)

Then press ENTER or =.

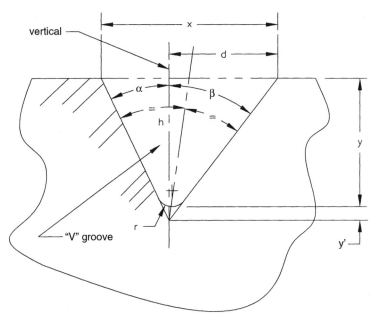

FIGURE 4.39 Solving plunge depth on angled notches.

The value for y' is calculated from the following equation:

$$y' = r\left\{\left[\frac{1}{\sin\left(\frac{\alpha+\beta}{2}\right)}\right]\cos\left(\frac{\beta-\alpha}{2}\right) - 1\right\} \qquad \text{(Eq. 4.2)}$$

Bracket the equation in the calculator as follows:

$r((1/\sin((\alpha+\beta)/2))(\cos((\beta-\alpha)/2)) - 1)$

Then press ENTER or =.
The distance d to the centerline of the cutter is calculated as follows:

$$\tan\beta = \frac{d}{(y+y')}$$

Then,

$$d = (y+y')\tan\beta \qquad \text{(Eq. 4.3)}$$

An actual problem is next shown in calculator entry form, following these basic equations.
Problem.

Given: $\alpha = 20°$, $\beta = 45°$, nose radius $r = 0.125$ in, groove width $x = 0.875$ in

4.34 CHAPTER FOUR

Find: Tool plunge distance y, distance y', and distance d from the preceding equations (see Fig. 4.39).

From Eq. 4.1:

$$y = (0.875 \cos 20° \cos 45°/\sin (20° + 45°)) - 0.125((1/\sin ((20° + 45°)/2))(\cos ((45° - 20°)/2)) - 1)$$

$$y = 0.5394 \text{ in}$$

From Eq. 4.2:

$$y' = 0.125((1/\sin ((20° + 45°)/2))(\cos ((45° - 20°)/2)) - 1)$$

$$y' = 0.1021 \text{ in}$$

From Eq. 4.3:

$$d = (0.5394 + 0.1021) \tan 45° \quad [\textbf{Note:}\ h = (y + y')]$$

$$d = (0.6415)\ 1.000$$

$$d = 0.6415 \text{ in}$$

Problem. A cutting tool with a nose radius r and angle θ is to cut a groove of x width. How deep is the plunge h from the surface of the work piece? (See Fig. 4.40.)

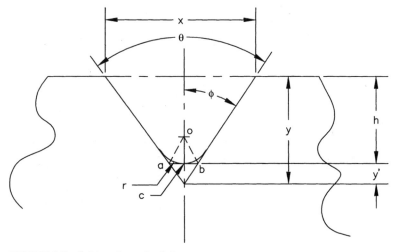

FIGURE 4.40 Solving plunge depth h.

Given: Width of groove $x = 0.875$ in, $\theta = 82°$, and $r = 0.125$ in

Step 1. Find distance *ab* from:

$$ab = 2r\left[\cot\left(\frac{90° + \phi}{2}\right)\right]$$

NOTE. $\phi = \dfrac{\theta}{2}$ or $ab = 2r\left[\dfrac{1}{\tan\left(\dfrac{90° + \phi}{2}\right)}\right]$

Step 2. Find y' from:

$$\tan\phi = \dfrac{cb}{y'}$$

$$y' = \dfrac{cb}{\tan\phi}$$

NOTE. $cb = \dfrac{ab}{2}$

Step 3. Find y from:

$$\tan\phi = \dfrac{x/2}{y}$$

$$y = \dfrac{x/2}{\tan\phi}$$

Step 4. Find h from:

$$h = y - y'$$

The solution to the preceding problem is numerically calculated as follows:

Step 1.

$$ab = 2(0.125)\,(\cot((90° + 41°)/2))$$
$$ab = 0.250\,(\cot 65.5°)$$
$$ab = 0.250\,(1/\tan 65.5°)$$
$$ab = 0.250 \times 0.4557$$
$$ab = 0.1139 \text{ in}$$

NOTE. $CB = ab/2$

Step 2.

$$y' = \dfrac{cb}{\tan\phi}$$
$$y' = (0.1139/2)/\tan 41°$$
$$y' = 0.0570/0.8693$$
$$y' = 0.0656 \text{ in}$$

Step 3.

$$y = (0.875/2)/\tan 41°$$

$$y = 0.4375/0.8693$$

$$y = 0.5033 \text{ in}$$

Step 4.

$$h = y - y'$$

$$h = 0.5033 - 0.0656$$

$$h = 0.4377 \text{ in}$$

Calculating and Checking V Grooves. See Fig. 4.41.

Problem. A V groove is to be machined to a width of 0.875 in, with an angle of 82°. Calculate the tool plunge depth y, and then check the width of the groove by calculating the height h that should be measured when a ball bearing of 0.500 in diameter is placed in the groove.

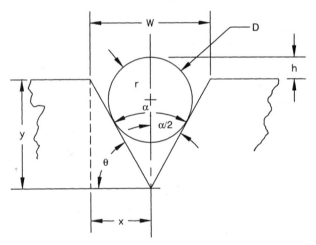

FIGURE 4.41 Checking groove width on angled notches.

Solution. Use the following two equations to calculate distances y and h:

Given: Groove width $W = 0.875$ in, groove angle $\alpha = 82°$

$$x = \frac{0.875}{2} = 0.4375$$

$$\theta = 90° - \frac{\alpha}{2}$$

MEASUREMENT AND CALCULATION PROCEDURES FOR MACHINISTS   4.37

$$\theta = 90° - \frac{82°}{2}$$

$$\theta = 90° - 41°$$

$$\theta = 49°$$

$$\tan \theta = \frac{y}{x} \qquad \text{(Eq. 4.4)}$$

$$y = x \tan \theta$$

$$y = 0.4375 \times \tan 49°$$

$$y = 0.4375 \times 1.1504$$

$$y = 0.5033 \text{ in depth of tool plunge}$$

Height h is calculated from the following equation, which is to be transposed for solving h:

$$W = 2 \tan \frac{\alpha}{2} \left(r \csc \frac{\alpha}{2} + r - h \right) \qquad \text{(Eq. 4.5)}$$

$$0.875 = 2 \tan \frac{82°}{2} \left(r \csc \frac{\alpha}{2} + 0.250 - h \right) \qquad \text{(Transpose this equation for } h.\text{)}$$

$$0.875 = 2 \tan 41°(r \csc 41° + 0.250 - h)$$

$$0.875 = 2 \times 0.8693 \left[r \left(\frac{1}{\sin 41°} \right) + 0.250 - h \right]$$

$$0.875 = 1.7386 \left[0.250 \left(\frac{1}{0.6561} \right) + 0.250 - h \right]$$

$$0.875 = 1.7386 \left[0.250(1.5242) + 0.250 - h \right]$$

$$0.875 = 1.7386(0.3811 + 0.250 - h)$$

$$0.875 = 0.6626 + 0.4347 - 1.7386h$$

$$1.7386h = 0.6626 + 0.4347 - 0.875$$

$$1.7386h = 0.2223$$

$$h = \frac{0.2223}{1.7368}$$

$$h = 0.1280 \text{ in}$$

NOTE. In the preceding equation, csc $\alpha/2$ was replaced with $1/(\sin \alpha/2)$, which is its equivalent. The reason for this substitution is that the cosecant function cannot be directly calculated on the pocket calculator. Since the cosecant, secant, and cotangent are equal to the reciprocals of the sine, cosine, and tangent, respectively, this substitution must be made, i.e., csc 41° = 1/sin 41°.

Arc Height Calculations. Figure 4.42 shows a method for finding the height h if an arc of known radius R is drawn tangent to two lines that are at a known angle A to each other. The simple equation for calculating h is given as follows:

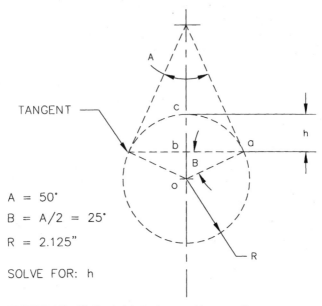

FIGURE 4.42 Finding height h of an arc of known radius.

$$bc = h = R\left(1 - \sin\frac{A}{2}\right); \quad \frac{A}{2} = B = 25°$$

where $A = 50°$
       $R = 2.125$ in

Therefore,

$$h = R(1 - \sin B)$$
$$h = 2.125(1 - \sin 25°)$$
$$h = 2.125(1 - 0.42262)$$
$$h = 2.125 \times 0.57738$$
$$h = 1.2269$$

Calculating Radii and Diameters Using Rollers or Pins. To calculate an inside radius or arc, see Fig. 4.43, and use the following equation:

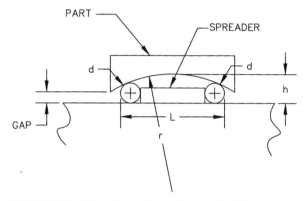

FIGURE 4.43 Calculating radius and diameter (inside).

Given: $d = 0.750$-in rollers or pins, $h = 1.765$ in measured, and $L = 10.688$ in measured

$$r = \frac{(L-d)^2}{8(h-d)} + \frac{h}{2}$$

$$r = \frac{(10.688 - 0.750)^2}{8(1.765 - 0.750)} + \frac{1.1765}{2}$$

$$r = \frac{98.7638}{8.120} + 0.8825$$

$$r = 12.1636 + 0.8825$$

$$r = 13.046 \text{ in}$$

To calculate an outside radius, diameter, or arc, see Fig. 4.44, and use the following equations:

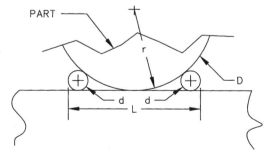

FIGURE 4.44 Calculating radius and diameter (outside).

Given: $L = 10.688$ in, $d = 0.750$ in; calculate r and D.

$$r = \frac{(L-d)^2}{8d}$$

$$r = \frac{(10.688 - 0.750)^2}{8(0.750)}$$

$$r = \frac{98.7638}{6}$$

$$r = 16.461 \text{ in}$$

$$D = \frac{(10.688 - 0.750)^2}{4(0.750)}$$

$$D = \frac{98.7638}{3}$$

$$D = 32.921$$

Calculating Blending Radius to Existing Arc. See Fig. 4.45.

Problem. Calculate the blending radius R_2 that is tangent to a given arc of radius R_1.

Solution. Distances X and Y and radius R_1 are known. Find radius R_2 when $X = 2.575$ in, $Y = 4.125$ in, and $R_1 = 5.198$ in.

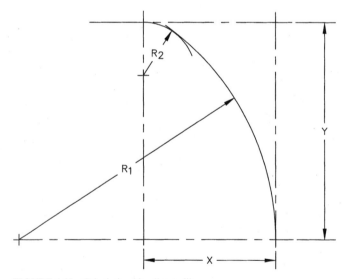

FIGURE 4.45 Calculating blending radii.

Use the following equation to solve for R_2:

$$R_2 = \frac{X^2 + Y^2 - 2R_1X}{2Y - 2R_1}$$

$$R_2 = \frac{(2.575)^2 + (4.125)^2 - 2(5.198)(2.575)}{2(4.125) - 2(5.198)}$$

$$R_2 = \frac{6.6306 + 17.0156 - 26.770}{8.250 - 10.396}$$

$$R_2 = \frac{-3.124}{-2.146}$$

$$R_2 = 1.456 \text{ in}$$

Plunge Depth of Milling Cutter for Keyways. See Fig. 4.46.

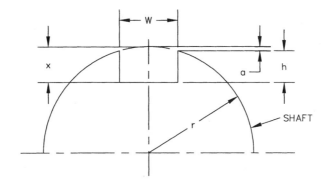

$a = x - h$
$r = \text{radius of shaft}$
$W = 0.250$
$h = 0.125$

FIGURE 4.46 Keyway depth, calculating.

EXAMPLE. Find the depth x the milling cutter must be sunk from the radial surface of the shaft to cut a shaft keyway with a width W of 0.250 in and a depth h of 0.125 in.
Given: $W = 0.250$ in, $h = 0.125$ in, $r = 0.500$ in (shaft diameter = 1.000 in)
Using the following equation, find the cutter plunge dimension x:

$$x = h + r - \sqrt{r^2 - \frac{W^2}{4}}$$

$$x = 0.125 + 0.500 - \sqrt{0.500^2 - \frac{0.250^2}{4}}$$

$$x = 0.125 + 0.500 - \sqrt{0.234375}$$

$$x = 0.625 - 0.484$$

$$x = 0.141 \text{ in}$$

From the figure, $a = x - h = 0.141 - 0.125 = 0.016$ in (reference dimension).

Keyway Cutting Dimensions. See Fig. 4.47 for calculation procedures.

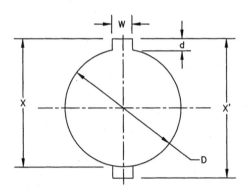

$$X = \sqrt{(D/d)^2 - (W/2)^2} + d + D/2$$

$$X' = 2X - D$$

FIGURE 4.47 Keyway cutting dimensions.

Compound Trigonometric Problem. In Fig. 4.48, we will solve for sides a and a' and length D, the distance from point 1 to point 2.

For side a, use the law of sines:

$$\frac{\sin B}{12} = \frac{\sin A}{a}$$

$$\frac{\sin 63°}{12} = \frac{\sin 54°}{a}$$

$$a = \frac{12(\sin 54°)}{\sin 63°}$$

$$a = \frac{12(0.809)}{0.891} = \frac{9.708}{0.891} = 10.896$$

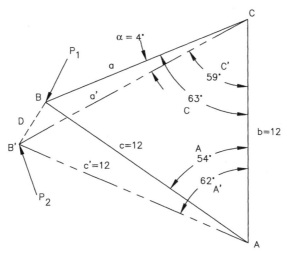

FIGURE 4.48 Compound trigonometric calculations. Note that angles A, B, and C form an isosceles triangle, as do angles A', B', and C'. Sides b, c, and $c' = 12$. When arm b moves from an angle of 54° to 62°, find the lengths of sides a and a', and the distance D from B to B' (P_1 to P_2).

For side a', also use the law of sines:

$$\frac{\sin B'}{b} = \frac{\sin A'}{a'}$$

$$\frac{\sin 59°}{12} = \frac{\sin 62°}{a'}$$

$$a' = \frac{12(\sin 62°)}{\sin 59°}$$

$$a' = \frac{12(0.883)}{0.857} = \frac{10.596}{0.857} = 12.364$$

For distance D, $(P_1 - P_2)$, use the law of cosines ($\alpha = 4°$, $a = 10.896$, $a' = 12.364$):

$$D^2 = (a)^2 + (a')^2 - 2(a)(a') \cos \alpha$$

$$D^2 = (10.896)^2 + (12.364)^2 - 2(10.896)(12.364)0.998$$

$$D^2 = 2.811$$

$$D = \sqrt{2.811} = 1.677 \quad \text{(distance between } P_1 \text{ and } P_2\text{)}$$

If sides a, a', and D are known, angle α can be calculated by transposing the law of cosines (see Fig. 4.48):

$$D^2 = (a)^2 + (a')^2 - 2(a)(a') \cos \alpha$$

$$2(a)(a') \cos \alpha = (a)^2 + (a')^2 - D^2$$

$$\cos \alpha = \frac{(a)^2 + (a')^2 - D^2}{2(a)(a')}$$

$$\cos \alpha = \frac{(10.896)^2 + (12.364)^2 - (1.677)^2}{2(10.896)(12.364)}$$

$$\cos \alpha = \frac{268.778983}{269.436288} = 0.997560$$

$$\arccos \alpha = 4.003° \quad (\text{accuracy} = 11'')$$

(If more accuracy is required, sides a, a', and D should be calculated to 6 decimal places.)

NOTE. Triangle C, B, B' can be checked with the Molleweide equation, after the other two angles are solved using the law of sines (see Chap. 1).

Transposing the Law of Cosines to Solve for the Angle

$$c^2 = a^2 + b^2 - 2ab \cos C$$

$$2ab \cos C = a^2 + b^2 - c^2 \quad (\text{rearranging})$$

$$\cos C = \frac{a^2 + b^2 - c^2}{2ab} \quad (\text{transposed})$$

Then take arccos C to find the angle C.
Transpose as shown to find cos A and cos B from:

$$a^2 = b^2 + c^2 - 2bc \cos A \quad \text{and} \quad b^2 = a^2 + c^2 - 2ac \cos B$$

Solving Heights of Triangles. From Fig. 4.49, if angle $A = 28°$, angle $C = 120°$, angle $C' = 60°$, and side $b = 14$ in, find the height X.

$$X = 14 \left(\frac{\sin 28° \sin 120°}{\sin (60° - 28°)} \right)$$

$$X = 14 \left(\frac{0.46947 \times 0.86603}{\sin 32°} \right) = 14 \left(\frac{0.40658}{0.52992} \right)$$

$$X = 14(0.76725) = 10.7415$$

Or, we can use:

MEASUREMENT AND CALCULATION PROCEDURES FOR MACHINISTS   4.45

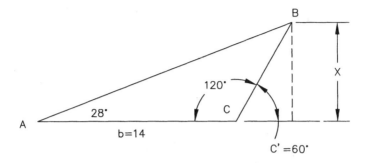

$$X = b\frac{\sin A \sin C}{\sin(C'-A)} \quad \text{or} \quad X = \frac{b}{\cot A - \cot C'}$$

FIGURE 4.49   Solving heights of oblique triangles.

$$X = \frac{14}{(1/\tan A) - (1/\tan C')} \quad \text{from} \quad \frac{b}{\cot A - \cot C'}$$

$$X = \frac{14}{1.88073 - 0.57735}$$

$$X = 10.7413$$

NOTE. $1/\tan A = \cot A$
Both equations check within 0.0002 in. If you need more accuracy, use more decimal places in the variables.
For oblique triangles, where no angle is greater than 90°, use the equations from Chap. 1 shown in the text.

Calculations Involving Properties of the Circle.   These include finding arc length, chord length, maximum height b, and the x, y ordinates. Refer to Fig. 4.50.

Given: Angle $\Theta = 42°$, radius $r = 6.250$ in

Find: Arc length ℓ, chord length c, maximum height b, height y when $x = 1.625$, and length x when $y = 0.125$.

$$\ell = \frac{\pi r \Theta°}{180} \qquad\qquad c = 2r \sin \frac{\theta}{2}$$

$$\ell = \frac{3.1416(6.250)42}{180} \qquad\qquad c = 2(6.250) \sin 21$$

$$\ell = \frac{824.668}{180} \qquad\qquad c = 2(6.250)0.3584$$

$$\ell = 4.581 \text{ in} \qquad\qquad c = 4.480 \text{ in}$$

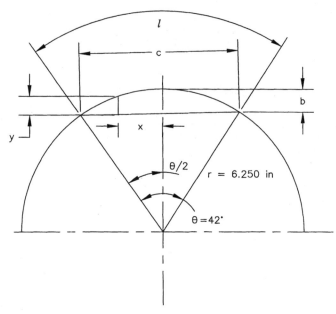

FIGURE 4.50 Calculation using properties of the circle.

$$b = \frac{c}{2}\left(\tan \frac{\theta}{4}\right) \quad \text{from} \quad y = b - r + \sqrt{r^2 - x^2}$$

$$b = \frac{4.480}{2}\left(\tan \frac{42}{4}\right) \quad\quad y = 0.4151 - 6.250 + \sqrt{6.250^2 - 1.625^2}$$

$$b = 2.240(\tan 10.5) \quad\quad y = 0.4151 - 6.250 + \sqrt{36.4219}$$

$$b = 2.240(0.1853) \quad\quad y = 0.4151 - 6.250 + 6.0351$$

$$b = 0.4151 \text{ in} \quad\quad y = 0.2002 \text{ in}$$

Find x when $y = 0.125$ in.

$$x = \sqrt{r^2 - (r + y - b)^2}$$

$$x = \sqrt{6.250^2 - (6.250 + 0.125 - 0.4151)}$$

$$x = \sqrt{3.5421}$$

$$x = 1.8820 \text{ in}$$

Using Simple Algebra to Solve Dimension-Scaling Problems. In Fig. 4.51, we have a scale drawing that has been reduced in size, such that the dimensions are not to actual scale. If we want to find a missing dimension, such as x in Fig. 4.51, we can ascertain the missing dimension using the simple proportion $a/b = c/d$, as follows:

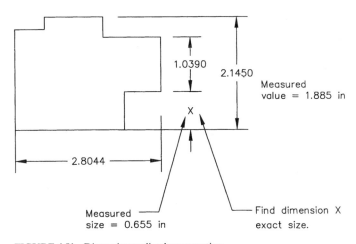

FIGURE 4.51 Dimension scaling by proportion.

The dimension 2.1450 was measured on the drawing as 1.885 in, and the missing dimension was measured on the drawing as 0.655 in. Therefore, a and c are the measured dimensions; b and x are the actual sizes. $d = x$.

$$\frac{a}{b} = \frac{c}{d}$$

$$\frac{1.885}{2.1450} = \frac{0.655}{x}$$

$$1.885x = 2.1450(0.655)$$

$$x = \frac{2.1450(0.655)}{1.885}$$

$$x = \frac{1.405}{1.885} = 0.745$$

Therefore, 0.745 in is the actual size of the missing dimension. This procedure is useful, but is only as accurate as the drawing and the measurements taken on the drawing. This procedure can also be used on objects in photographs that do not have perspective distortion, where one aspect or dimensional feature is known and can be measured.

Useful Geometric Proportions. In reference to Fig. 4.52, when ab is the diameter, and dc is a perpendicular line drawn from the diameter that intersects the circle, the following proportion is valid:

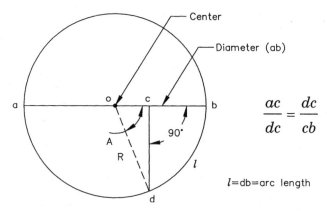

$$\frac{ac}{dc} = \frac{dc}{cb}$$

$l = db =$ arc length

FIGURE 4.52 Proportion problem in the circle.

$$\frac{ac}{dc} = \frac{dc}{cb}$$

If $ac = 6$ and $dc = 5$, find the length cb.

$$\frac{6}{5} = \frac{5}{cb}$$

$$6cb = 25$$

$$cb = \frac{25}{6} = 4.167 \text{ in}$$

The diameter ab is then:

$$ab = 6 + 4.167$$

$$ab = 10.167 \text{ in}$$

$$R = \frac{10.167}{2} = 5.084 \text{ in} \quad \text{(radius)}$$

The internal angle of the arc db can be calculated by finding the length oc:

$$oc = R - cb$$

$$oc = 5.084 - 4.167 = 0.917 \text{ in}$$

and then solving the right triangle ocd:

$$\tan A = \frac{dc}{oc}$$

$$\tan A = \frac{5}{0.917} = 5.4526$$

arctan 5.4526 = 79.6075°

angle A = 79.6075°

The arc length db can then be calculated from the properties of the circle:

$$db = \frac{\pi RA}{180} = \ell$$

$$\ell = \frac{(3.1416)(5.084)(79.6075)}{180}$$

$$\ell = \frac{1271.483}{180} = 7.064 \text{ in}$$

The sum of all the internal angles of any polygon (Fig. 4.53) is equal to the number of sides minus 2, times 180°:

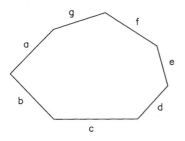

FIGURE 4.53 Polygon.

$a + b + c + d + e + f + g = (7 - 2) \times 180°$

$5(180°) = 900°$ (sum of internal angles)

In any triangle, a straight line drawn between two sides, which is parallel to the third side, divides those sides proportionally (see Fig. 4.54). Therefore:

$$\frac{Ad}{dB} = \frac{Ae}{eC}$$

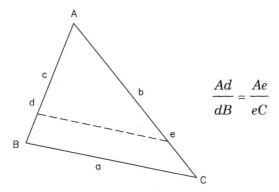

$$\frac{Ad}{dB} = \frac{Ae}{eC}$$

FIGURE 4.54 Proportions in triangles.

EXAMPLE. If Ad = 4 in, dB = 1 in, and Ae = 6 in, find eC:

$$\frac{4}{1} = \frac{6}{eC}$$

$$4eC = 6$$

$$eC = \frac{6}{4} = 1.5 \text{ in}$$

In the same triangle, the following proportions are also true:

$$\frac{Ad}{AB} = \frac{de}{BC} \quad \text{and} \quad \frac{Ae}{AC} = \frac{de}{BC}$$

Proof of the Proportions Shown in Fig. 4.54. If angle $A = 50°$, solve the triangle for side BC.

From the law of cosines (units in degrees and inches):

$$a = BC$$

$$c = Ad + dB = 4 + 1 = 5$$

$$b = Ae + eC = 6 + 1.5 = 7.5$$

Then:

$$a^2 = b^2 + c^2 - 2bc \cos 50°$$

$$a^2 = (7.5)^2 + (5)^2 - 2(5 \times 7.5)0.64278$$

$$a = \sqrt{33.0409}$$

$$a = 5.748$$

Solving side de by the law of cosines, $de = 4.5985$. Therefore:

$$\frac{Ad}{AB} = \frac{de}{BC} \quad AB = Ad + dB = 5$$

$$de = \frac{Ad \times BC}{AB}$$

$$de = \frac{4 \times 5.748}{5}$$

$$de = 4.5984 \quad \text{(Proof of the proportion)}$$

Lengths of circular arcs with the same center angle are proportional to the lengths of the radii (see Fig. 4.55).

EXAMPLE. If $a = 2.125$, $r = 3$, and $R = 4.250$, find arc length b.

$$\frac{a}{b} = \frac{r}{R}$$

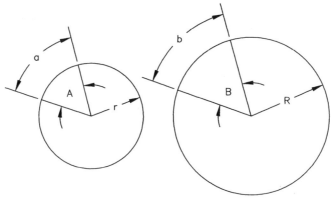

If angle A = angle B

$$\frac{a}{b} = \frac{r}{R}$$

FIGURE 4.55 Lengths of circular arcs.

$$\frac{2.125}{b} = \frac{3}{4.25}$$

$$3b = 9.031225$$

$$b = \frac{9.03125}{3} = 3.0104$$

Sample Trigonometry Problem. See Fig. 4.56.
Problem. The dimensions of three sides of a triangle are known.
Find: Altitude x, and the location of x by dimensions y and z.

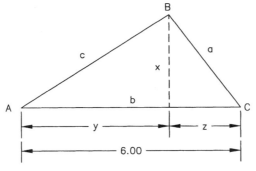

FIGURE 4.56 Solving the oblique triangle.

First, find angles C and A from the law of cosines and then the law of sines. Finding angle C:

$$c^2 = a^2 + b^2 - 2ab \cos C$$

$$\cos C = \frac{a^2 + b^2 - c^2}{2ab}$$

$$\cos C = \frac{(4.17)^2 + (6)^2 - (5.45)^2}{2(4.17 \times 6)}$$

$$\cos C = \frac{23.6864}{50.04} = 0.473349$$

$$\arccos 0.473349 = 61.7481°$$

$$\text{angle } C = 61.7481°$$

Find angle A from the law of sines:

$$\frac{\sin A}{a} = \frac{\sin C}{c}$$

$$\sin A = \frac{a \sin C}{c}$$

$$\sin A = \frac{4.17 \sin 61.7481}{5.45} = \frac{3.67325}{5.45} = 0.67399$$

$$\arcsin 0.67399 = 42.3758°$$

$$\text{angle } A = 42.3758°$$

Now, angle $B = 180° - (A + C)$:

$$B = 180° - (42.3758 + 61.7481)$$

$$B = 180° - 104.1239°$$

$$\text{angle } B = 75.8761°$$

Now, solve for altitude x (see previous calculations for angles A and C):

$$x = b \frac{\sin A \sin C}{\sin (A + C)}$$

$$x = b \left(\frac{0.67399 \times 0.473349}{\sin (42.3758 + 61.7481)} \right)$$

$$x = 6 \left(\frac{0.31903}{0.96977} \right)$$

$$x = 6(0.32897)$$

$$x = 1.974 \text{ in}$$

MEASUREMENT AND CALCULATION PROCEDURES FOR MACHINISTS   **4.53**

Now, find y and z:

$$\tan C = \frac{x}{z}$$

$$z = \frac{x}{\tan C} = \frac{1.974}{\tan 61.7481}$$

$$z = \frac{1.974}{1.861}$$

$$z = 1.061 \text{ in}$$

Since $y = b - z$ and $b = 6$,

$$y = 6 - 1.061 = 4.939 \text{ in}$$

Sample Countersinking Problem.   See Fig. 4.57.
Problem.   What is the diameter of the countersink D when we want the head of the flathead bolt or screw to be 0.010 in below the surface of the part? (See Fig. 4.57a.)

Given: Head diameter of an 82°, 0.250-in-diameter flathead screw Hd = 0.740 in; depth of head below the surface of the part $x = 0.010$ in.

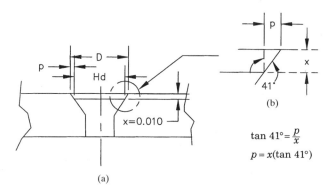

FIGURE 4.57   Countersinking calculations.

Solve the right triangle shown in Fig. 4.57b, for side p:

$$\tan 41° = \frac{p}{x}$$

$$p = x(\tan 41°)$$

$$p = 0.010(0.8693)$$

$$p = 0.00869 \text{ in}$$

4.54 CHAPTER FOUR

Then, the final diameter of the countersink D is found:

$$D = Hd + 2(p)$$

$$D = 0.740 + 2(0.00869)$$

$$D = 0.740 + 0.017$$

$$D = 0.757 \text{ in}$$

NOTE. Measure the diameter of the head of the screw or bolt Hd with a micrometer prior to doing the calculations. Different manufacturers produce different head diameters on flathead screws or bolts, according to the tolerances allowed by ANSI standards for fasteners. The diameter of 0.740 in used in the preceding problem is an average value.

4.4 FINDING COMPLEX ANGLES FOR MACHINED SURFACES

Compound Angle Problems. Figure 4.58 shows a quadrangular pyramid with four right-angle triangles as sides and a rectangular base, $OBCD$.

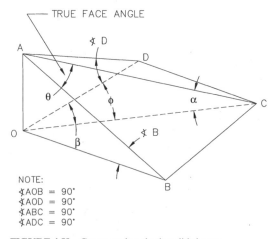

FIGURE 4.58 Compound angles in solid shapes.

Problem. If a plane is passed through AOC, find the compound angles α, β, and ϕ when angle B and angle D are known.

Given: Angle $B = 24°$, angle $D = 25°$.

MEASUREMENT AND CALCULATION PROCEDURES FOR MACHINISTS

Solution. From the compound angle relations shown in Fig. 4.2, the following equations are used to find angles β, ϕ, and α:

$$\tan \beta = \tan B \cot D \quad \text{(Eq. 4.6)}$$

$$\tan \phi = \cot B \tan D \quad \text{(Eq. 4.7)}$$

$$\cot \alpha = \sqrt{\cot^2 B + \cot^2 D} \quad \text{(Eq. 4.8)}$$

Solving for angle β (from Eq. 4.6):

$$\tan \beta = \tan 24° \times \cot 35°$$

$$\tan \beta = 0.4452 \times \left(\frac{1}{\tan 35°}\right)$$

$$\tan \beta = 0.4452 \times 1.4281$$

$$\tan \beta = 0.6358$$

$$\arctan 0.6358 = 32.448° = \text{angle } \beta$$

Solving for angle ϕ (from Eq. 4.7):

$$\tan \phi = \cot 24° \times \tan 35°$$

$$\tan \phi = \left(\frac{1}{\tan 24°}\right) \times \tan 35°$$

$$\tan \phi = 2.2460 \times 0.7002$$

$$\tan \phi = 1.5726$$

$$\arctan 1.5726 = 57.548° = \text{angle } \phi$$

Solving for angle α (from Eq. 4.8):

$$\cot \alpha = \sqrt{\cot^2 B + \cot^2 D}$$

$$\cot \alpha = \sqrt{(2.2460)^2 + (1.4281)^2}$$

$$\cot \alpha = \sqrt{7.084}$$

$$\cot \alpha = 2.6615$$

$$\tan \alpha = \frac{1}{2.6615}$$

$$\tan \alpha = 0.3757$$

$$\arctan 0.3757 = 20.591° = \text{angle } \alpha$$

Problem. Find the true face angle θ.

Given: Side $OB = 6.000$ in.

Solution. First, calculate the length of side OA:

$$\tan B = \frac{OA}{OB}$$

$$\tan 24° = \frac{OA}{6.000}$$

$$OA = 6.000 \tan 24°$$

$$OA = 6.000 \times 0.4452$$

$$OA = 2.6712$$

Next, calculate the length OD (note that length $BC = OD$):

$$\tan D = \frac{OA}{OD}$$

$$OD = \frac{OA}{\tan D}$$

$$OD = \frac{2.6712}{\tan 35°}$$

$$OD = \frac{2.6712}{0.7002}$$

$$OD = 3.8149 \text{ in}$$

Problem. Find the true face angle θ.
Solution. First, calculate the length of side AB:

$$\cos B = \frac{OB}{AB}$$

$$\cos 24° = \frac{6.000}{AB}$$

$$AB = \frac{6.000}{\cos 24°}$$

$$AB = \frac{6.000}{0.9135}$$

$$AB = 6.5681$$

Next, calculate the length of side AC from the pythagorean theorem:

$$AC^2 = AB^2 + BC^2$$

$$AC = \sqrt{(6.5681)^2 + (3.8152)^2}$$

$$AC = \sqrt{57.6957}$$

$$AC = 7.5958 \text{ in}$$

Then, calculate the face angle θ from the law of cosines:

$$BC^2 = AB^2 + AC^2 - 2(AB)(AC) \cos\theta$$

$$(3.8152)^2 = (6.5681)^2 + (7.5958)^2 - 2(6.5681)(7.5958) \cos\theta$$

$$14.5558 = 43.1399 + 57.6962 - 99.7799 \cos\theta$$

$$99.7799 \cos\theta = 43.1399 + 57.6962 - 14.5558$$

$$\cos\theta = \frac{86.2803}{99.7799}$$

$$\cos\theta = 0.8647$$

$$\text{arccos } 0.8647 = 30.152° = \text{angle } \theta$$

The true face angle θ is therefore 30.152°.
Problem. Check angle ABC for a right triangle.
Solution. Solve angle ACB (note that $BC = OD = 3.8149$ in):

$$\tan \measuredangle ACB = \frac{AB}{BC}$$

$$\tan \measuredangle ACB = \frac{6.5681}{3.8149}$$

$$\tan \measuredangle ACB = 1.7217$$

$$\text{arctan } 1.7217 = 59.851°$$

Therefore,

$$\theta + 59.851° + 90° = 30.152° + 59.851° + 90° = 180.003°$$

This indicates that the calculated angles ACB and θ are accurate within 0.003° or 0.18′ of arc. Using more decimal places for the calculated sides and angles will produce more accurate results, if required.

Compound Angle Problem—Milling an Angled Plane. See Fig. 4.59.
Problem. A rectangular block, shown in Fig. 4.59, is milled off to form a triangular plane ABC, and the angles formed by the edges of the rectangular plane to the bottom of the block are known. Calculate the compound angle θ; sides *a, b,* and *c;* and angles *A* and *B*.

Given: Length of block = 5.250 in, width = 3.750 in, and height = 2.500 in; angles $\alpha = 23°$ and $\beta = 33°$; $h = 0.625$ in; and $h' = 2.500 - 0.625 = 1.875$ in.

4.58 CHAPTER FOUR

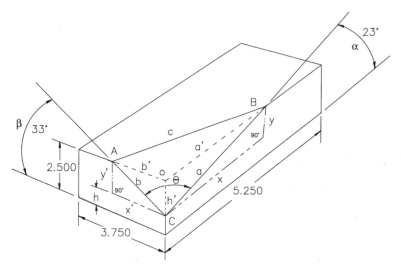

FIGURE 4.59 Milling an angular plane, problem.

Solution. Use the compound angle equation for angle θ (note that angle θ = angle *C*):

$$\cos \theta = \sin \alpha \sin \beta$$

$$\cos \theta = \sin 23° \times \sin 33°$$

$$\cos \theta = 0.39073 \times 0.54464$$

$$\cos \theta = 0.21821$$

$$\arccos 0.21821 = 77.7129° = \text{angle } \theta$$

Calculate side *a*:

$$\sin 23° = \frac{y}{a}$$

$$a = \frac{y}{\sin 23°}$$

$$a = \frac{1.875}{0.39073}$$

$$a = 4.7987 \text{ in}$$

Calculate side b:

$$\sin 33° = \frac{y'}{b} \qquad y' = 2.500 - 0.625 = 1.875$$

$$b = \frac{y'}{\sin 33°}$$

$$b = \frac{1.875}{0.5446}$$

$$b = 3.4429 \text{ in}$$

Now, calculate side c using the law of cosines:

$$c^2 = a^2 + b^2 - 2ab \cos \theta$$

$$c = \sqrt{(4.7987)^2 + (3.4426)^2 - 2(4.7987)(3.4426)0.21281}$$

$$c = \sqrt{27.8478}$$

$$c = 5.2771 \text{ in}$$

Calculate angle A using the law of sines:

$$\frac{c}{\sin \theta} = \frac{a}{\sin A}$$

$$\sin A = \frac{a \sin \theta}{c}$$

$$\sin A = \frac{4.7987 \times 0.9771}{5.2271} = 0.8970$$

$$\arcsin 0.8970 = 63.7665° = \text{angle } A$$

Calculate angle B using the law of sines:

$$\frac{c}{\sin \theta} = \frac{b}{\sin B}$$

$$\sin B = \frac{b \sin \theta}{c}$$

$$\sin B = \frac{3.4426 \times 0.9771}{5.2771} = 0.6374$$

$$\arcsin 0.6374 = 39.5982° = \text{angle } B$$

Now, check the sum of the angles in triangle *ABC*. *Rule:* The sum of the angles in any triangle must equal 180°. Therefore:

$$\angle A + \angle B + \angle C = 180°$$
$$62.6879° + 39.5982° + 77.7129° = 180°$$
$$179.999° = 180°$$

The calculated angles check within 0.001°, which is within 0.06′ or 3.6″.

To find the volume of material removed from the block, use the following equation and the other distances, a', b', and h', which can be easily calculated, as shown in Fig. 4.59.

$$V = \frac{1}{3}\left[\left(\frac{b' \times h'}{2}\right) \times a'\right]$$

Sample Problems for Calculating Compound Angles in Three-Dimensional Parts. Referring to Table 4.1, we will find angle γ when we know angles α and β, using the following equation:

$$\cos \gamma = \frac{\tan \beta}{\tan \alpha}$$

TABLE 4.1 Trigonometric Relations for Compound Angles (See Fig. 4.60)

Given	To find	Equation
α and β	γ	$\cos \gamma = \dfrac{\tan \beta}{\tan \alpha}$
α and β	δ	$\cos \delta = \dfrac{\sin \beta}{\sin \alpha}$
α and γ	β	$\tan \beta = \cos \gamma \tan \alpha$
α and γ	δ	$\tan \delta = \cos \alpha \tan \gamma$
α and δ	β	$\sin \beta = \sin \alpha \cos \delta$
α and δ	γ	$\tan \gamma = \dfrac{\tan \delta}{\cos \alpha}$
β and γ	α	$\tan \alpha = \dfrac{\tan \beta}{\cos \gamma}$
β and γ	δ	$\sin \delta = \cos \beta \sin \gamma$
β and δ	α	$\sin \alpha = \dfrac{\sin \beta}{\cos \delta}$
β and δ	γ	$\sin \gamma = \dfrac{\sin \delta}{\cos \beta}$
γ and δ	α	$\cos \alpha = \dfrac{\tan \delta}{\tan \gamma}$
γ and δ	β	$\cos \beta = \dfrac{\sin \delta}{\sin \gamma}$

NOTE. In Fig. 4.60, the corner angles marked with a box are 90° right angles. To solve the problem, we must first calculate angles α and β.
Solution. We must know or measure the distances *ov*, *om*, and *mn*. If *ov* = 2.125 in, *om* = 4.875 in, and *mn* = 6.500 in, first find angle α:

$$\tan \alpha = \frac{ov}{om} = \frac{2.125}{4.875}$$

$$\tan \alpha = 0.435897$$

$$\arctan 0.435897 = 23.5523° = \text{angle } \alpha$$

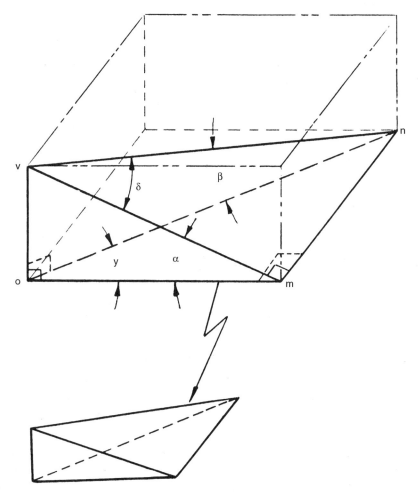

FIGURE 4.60 Calculating compound angles.

To find angle β, we must first find the diagonal length *on:*

$$on^2 = om^2 + mn^2 \quad \text{(where } om \text{ and } mn \text{ are known)}$$

$$on^2 = (4.875)^2 + (6.500)^2$$

$$on^2 = 66.015625$$

$$on = \sqrt{66.015625}$$

$$on = 8.125 \text{ in}$$

Then, find angle β:

$$\tan \beta = \frac{ov}{on} = \frac{2.125}{8.125} = 0.261538$$

$$\arctan 0.261538 = 14.656751° = \text{angle } \beta$$

We now know angles α and β, and we can find angle γ using the equation from Table 4.1:

$$\cos \gamma = \frac{\tan \beta}{\tan \alpha}$$

where tan β = 0.261538 (from previous calculation)
tan α = 0.435897 (from previous calculation)

Then,

$$\cos \gamma = \frac{0.261538}{0.435897}$$

$$\cos \gamma = 0.599999$$

$$\arccos 0.599999 = 53.130174° = \text{angle } \gamma$$

Problem. Prove the following relationship from Table 4.1:

$$\cos \delta = \frac{\sin \beta}{\sin \alpha}$$

Solution. First, find the length of the diagonal *vm:*

$$vm^2 = ov^2 + om^2$$

$$vm^2 = (2.125)^2 + (4.875)^2$$

$$vm^2 = 28.28125$$

$$vm = \sqrt{28.28125}$$

$$vm = 5.318012 \text{ in}$$

MEASUREMENT AND CALCULATION PROCEDURES FOR MACHINISTS   **4.63**

Then, calculate angle δ:

$$\tan \delta = \frac{mn}{vm}$$

$$\tan \delta = \frac{6.500}{5.318012} = 1.222261$$

$$\arctan 1.222261 = 50.711490° = \text{angle } \delta$$

Then, use the equation from Table 4.1 to see if angle $\delta = 50.711490°$:

$$\cos \delta = \frac{\sin \beta}{\sin \alpha}$$

$$\cos \delta = \frac{\sin 14.656751°}{\sin 23.5523°}$$

$$\cos \delta = \frac{0.253028}{0.399586}$$

$$\cos \delta = 0.633225$$

$$\arccos 0.633225 = 50.71154° = \text{angle } \delta$$

Previously, we calculated angle $\delta = 50.71149°$. So, the relationship is valid. The accuracy of the preceding relationship, as calculated, is accurate to within $50.71154° - 50.71149° = 0.00005°$, or $0.18''$ of arc.

Also, from the relationship $\tan \beta = \cos \gamma \tan \alpha$, we will check angle β, which was previously calculated as $14.656751°$; angle $\alpha = 23.5523°$; and angle $\gamma = 53.130174°$, as follows:

$$\tan \beta = \cos 53.130174° \times \tan 23.5523°$$

$$\tan \beta = 0.261538$$

$$\arctan 0.261538 = 14.656726° = \text{angle } \beta$$

Angle β was previously calculated as $14.656751°$, which also checks within $14.656751 - 14.656726 = 0.000025°$, or $0.09''$ of arc.

The preceding calculations are useful in machining work and tool setup, and also show the validity of the angular and trigonometric relationships of compound angles on three-dimensional objects, as shown in Figs. 4.2 and 4.60 and Table 4.1.

CHAPTER 5
FORMULAS AND CALCULATIONS FOR MACHINING OPERATIONS

5.1 TURNING OPERATIONS

Metal removal from cylindrical parts is accomplished using standard types of engine lathes or modern machining centers, the latter operated by computer numerical control (CNC). Figure 5.1 shows a typical large geared-head engine lathe with a digital two-axis readout panel at the upper left of the machine. Figure 5.2a shows a modern high-speed CNC machining center. The machining center is capable of highly accurate and rapid production of machined parts. These modern machining centers are the counterparts of engine lathes, turret lathes, and automatic screw machines when the turned parts are within the capacity or rating of the machining center. Figure 5.2b shows a view of the CNC turning center's control panel.

Cutting Speed. Cutting speed is given in surface feet per minute (sfpm) and is the speed of the workpiece in relation to the stationary tool bit at the cutting point surface. The cutting speed is given by the simple relation

$$S = \frac{\pi d_f (\text{rpm})}{12} \quad \text{for inch units}$$

and

$$S = \frac{\pi d_f (\text{rpm})}{1000} \quad \text{for metric units}$$

where S = cutting speed, sfpm or m/min
 d_f = diameter of work, in or mm
 rpm = revolutions per minute of the workpiece

When the cutting speed (sfpm) is given for the material, the revolutions per minute (rpm) of the workpiece or lathe spindle can be found from

$$\text{rpm} = \frac{12S}{\pi d_f} \quad \text{for inch units}$$

FIGURE 5.1 A typical geared-head engine lathe.

FORMULAS AND CALCULATIONS FOR MACHINING OPERATIONS 5.3

FIGURE 5.2a A modern CNC turning center.

and $\qquad \text{rpm} = \dfrac{1000S}{\pi d_f} \qquad$ for metric units

EXAMPLE. A 2-in-diameter metal rod has an allowable cutting speed of 300 sfpm for a given depth of cut and feed. At what revolutions per minute (rpm) should the machine be set to rotate the work?

$$\text{rpm} = \frac{12S}{\pi d_f} = \frac{12(300)}{3.14 \times 2} = \frac{3600}{6.283} = 573 \text{ rpm}$$

Set the machine speed to the next closest lower speed that the machine is capable of attaining.

Lathe Cutting Time. The time required to make any particular cut on a lathe or turning center may be found using two methods. When the cutting speed is given, the following simple relation may be used:

$$T = \frac{\pi d_f L}{12FS} \qquad \text{for inch units}$$

and $\qquad T = \dfrac{\pi d_f L}{1000FS} \qquad$ for metric units

FIGURE 5.2b Turning center control panel.

where T = time for the cut, min
d_f = diameter of work, in or mm
L = length of cut, in or mm
F = feed, inches per revolution (ipr) or millimeters per revolution (mmpr)
S = cutting speed, sfpm or m/min

FORMULAS AND CALCULATIONS FOR MACHINING OPERATIONS   5.5

EXAMPLE. What is the cutting time in minutes for one pass over a 10-in length of 2.25-in-diameter rod when the cutting speed allowable is 250 sfpm with a feed of 0.03 ipr?

$$T = \frac{\pi d_f L}{12 FS} = \frac{3.1416(2.25)10}{12(0.03)250} = \frac{70.686}{90} = 0.785 \text{ min, or 47 sec}$$

When the speed in rpm of the machine spindle is known, the cutting time may be found from

$$T = \frac{L}{F(\text{rpm})}$$

where L = length of work, in
T = cutting time, min
F = feed, ipr
rpm = spindle speed or workpiece speed, rpm

Volume of Metal Removed. The volume of metal removed during a lathe cutting operation can be calculated as follows:

$$V_r = 12 C_d FS \quad \text{for inch units}$$

and $\quad V_r = C_d FS \quad$ for metric units

where V_r = volume of metal removed, in³/min or cm³/min
C_d = depth of cut, in or mm
F = feed, ipr or mmpr
S = cutting speed, sfpm or m/min

NOTE. 1 in³ = 16.387 cm³

EXAMPLE. With a depth of cut of 0.25 in and a feed of 0.125 in, what volume of material is removed in 1 min when the cutting speed is 120 sfpm?

$$V_r = 12 C_d FS = 12 \times 0.25 \times 0.125 \times 120 = 45 \text{ in}^3/\text{min}$$

For convenience, the chart shown in Fig. 5.3 may be used for quick calculations of volume of material removed for various depths of cut, feeds, and speeds.

Machine Power Requirements (Horsepower or Kilowatts). It is often necessary to know the machine power requirements for an anticipated feed, speed, and depth of cut for a particular material or class of materials to see if the machine is capable of sustaining the desired production rate. The following simple formulas for calculating required horsepower are approximate only because of the complex nature and many variables involved in cutting any material.

The following formula is for approximating machine power requirements for making a particular cut:

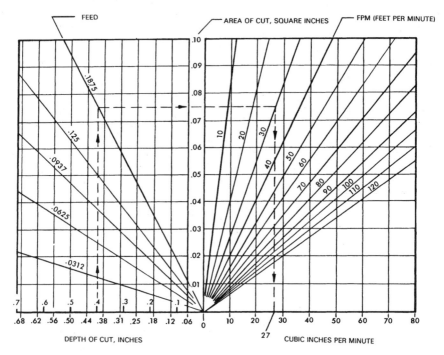

FIGURE 5.3 Metal-removal rate (mrr) chart.

$$hp = dfSC$$

where hp = required machine horsepower
 d = depth of cut, in
 f = feed, ipr
 S = cutting speed, sfpm
 C = power constant for the particular material (see Fig. 5.4)

EXAMPLE. With a depth of cut of 0.06 in and a feed of 0.025 in, what is the power requirement for turning aluminum-alloy bar stock at a speed of 350 sfpm?

$$hp = dfSC = 0.06 \times 0.025 \times 350 \times 4 \quad \text{(see Fig. 5.4)}$$

$$= 2.1 \text{ hp}$$

For the metric system, the kilowatt requirement is 2.1 hp × 0.746 kW/hp = 1.76 kW.

NOTE. 0.746 kW = 1 hp or 746 W = 1 hp.

The national manufacturers of cutting tools at one time provided the users of their materials with various devices for quickly approximating the various machining calculations shown in the preceding formulas. With the pocket calculator, these devices are no longer required, and the calculations are more accurate.

Power Constants for Various Metals and Alloys

Material	Constant
SAE Steels:	
1005 - 1029	6
1030 - 1050	7
1053 - 1095	8
1211 - 1215	6
1314 - 1345	6
1330 - 1350	9
1524 - 1552	9
4130 - 4820	9
5120 - 52100	10
Cast steels	9
SAE Stainless steels:	
30303, 51403, 51410, 51416 51431, 51430F, 51440F	10
30302, 30304, 30309, 30316 30321, 51431, 51501	11
51420, 51420F, 51440A, B, C	12
Cast irons:	
Hard	4
Medium	3
Soft	3
Semi-steel	3
Malleable irons:	
Hard	5
Medium	4
Soft	3

Material	Constant
Titanium & alloys:	
Pure	4
Alpha alloys	7
Beta alloys	8
Copper	4
Zinc alloys	3
Monel	12
Brass & bronze:	
Hard	10
Soft	4
Aluminum alloys:	
Cast	3
Bar stock	4
Magnesium alloys	3

FIGURE 5.4 Power constant table.

Although formulas and calculators are available for doing the various machining calculations, it is to be cautioned that these calculations are approximations and that the following factors must be taken into consideration when metals and other materials are cut at high powers and speeds using modern cutting tools.

1. Available machine power
2. Condition of the machine
3. Size, strength, and rigidity of the workpiece
4. Size, strength, and rigidity of the cutting tool

Prior to beginning a large production run of turned parts, sample pieces are run in order to determine the exact feeds and speeds required for a particular material and cutting tool combination.

Power Constants. Figure 5.4 shows a table of constants for various materials which may be used when calculating the approximate power requirements of the cutting machines.

Speeds, Cuts, and Feeds for Turning Operations

High-Speed Steel (HSS), Cast-Alloy, and Carbide Tools (See Fig. 5.5). The surface speed (sfpm), depth of cut (in), and feed (ipr) for various materials using high-speed steel (HSS), cast-alloy, and carbide cutting tools are shown in Fig. 5.5. In all cases, especially where combinations of values are selected that have not been used previously on a given machine, the selected values should have their required horsepower or kilowatts calculated. Use the approximate calculations shown previously, or use one of the machining calculators available from the cutting tool manufacturers. The method indicated earlier for calculating the required horsepower gives a conservative value that is higher than the actual power required. In any event, on a manually controlled machine, the machinist or machine operator will know if the selected speed, depth of cut, and feed are more than the given machine can tolerate and can make corrections accordingly. On computer numerically controlled and direct numerically controlled (CNC/DNC) automatic turning centers and other automatic machines, the cutting parameters must be selected carefully, with the machine operator carefully watching the first trial program run so that he or she may intervene if problems of overloading or tool damage occur.

Procedures for Selection of Speed, Feed, and Depth of Cut. Use the preceding speed, feed, and depth of cut figures as a basis for these choices. Useful tool life is influenced most by cutting speed. The feed rate is the next most influential factor in tool life, followed by the depth of cut (doc).

When the depth of cut exceeds approximately 10 times the feed rate, a further increase in depth of cut has little effect on tool life. In selecting the cutting conditions for a turning or boring operation, the first step is to select the depth of cut, followed by selection of the feed rate and then the cutting speed. Use the preceding horsepower/kilowatt equations to determine the approximate power requirements for a particular depth of cut, feed rate, and cutting speed to see if the machine can handle the power required.

Class	Material or SAE No.	Cutting Tool Material	Depth of Cut: 0.005-0.015 Feed: 0.002-0.005	Depth of Cut: 0.015-0.094 Feed: 0.005-0.015	Depth of Cut: 0.094-0.187 Feed: 0.015-0.030	Depth of Cut: 0.187-0.375 Feed: 0.030-0.050	Depth of Cut: 0.375-0.750 Feed: 0.050-0.090
Free-cutting steels	1112, 11L17 1120 1315, etc.	HSS Cast-alloys Sintered carbides	.... 750 - 1,500	250 - 350 425 - 550 600 - 750	175 - 250 315 - 400 450 - 600	80 - 150 215 - 300 350 - 450	55 - 75 100 - 210 175 - 350
Carbon & low-alloy steels	1010, 1020 1025, etc.	HSS Cast-alloys Sintered carbides	.... 700 - 1,200	225 - 300 375 - 500 550 - 700	150 - 200 275 - 350 400 - 550	75 - 125 180 - 250 300 - 400	45 - 65 100 - 175 150 - 300
Medium alloy steels	1030 1050	HSS Cast-alloys Sintered carbides	.... 600 - 1,000	200 - 275 325 - 400 450 - 600	125 - 175 230 - 300 350 - 450	70 - 120 160 - 225 250 - 350	40 - 60 80 - 150 125 - 250
High-alloy steels	1060 1095 1350	HSS Cast-alloys Sintered carbides	.... 500 - 750	175 - 250 250 - 350 400 - 500	125 - 175 200 - 250 300 - 400	65 - 100 150 - 200 200 - 300	35 - 55 65 - 150 100 -200
Chromium-nickel steels	3120, 3450 5140, 52100	HSS Cast-alloy Sintered carbides	.... 425 - 550	150 - 200 230 - 315 325 - 425	100 - 125 165 - 225 250 - 325	50 - 75 110 - 160 175 - 250	30 - 50 55 - 110 75 - 175
Molybdenum steels	4130 4615	HSS Cast-alloys Sintered carbides	.... 475 - 650	160 - 210 250 - 325 350 - 475	110 - 140 160 - 225 275 - 350	60 - 80 120 - 150 200 - 275	35 - 55 65 - 100 100 - 200
Chrome, vanadium and stainless steels	6120 6150, 18Cr-5Ni 6195	HSS Cast-alloys Sintered carbides	.... 375 - 500	100 - 150 210 - 250 300 - 375	80 - 100 170 - 200 250 - 300	50 - 75 110 - 165 175 - 250	30 - 50 55 - 100 75 - 175
Tungsten steels	7260 18-4-1 annealed	HSS Cast-alloys Sintered carbides	.... 325 - 400	120 - 150 130 - 175 250 - 325	75 - 120 110 - 130 200 - 250	40 - 75 80 - 100 150 - 200	25 - 40 35 - 80 50 - 150
Special steels	12-14% Mn.	HSS Cast-alloys Sintered carbides	.... 200 - 250	.... 125 - 200	.... 75 - 125	.... 50 - 75	....

FIGURE 5.5 Cuts, feeds, and speeds table.

Material	Tool						
Special steels Si. elect., sheet ingot iron, etc.	HSS Cast-alloys Sintered carbide	400 - 500 1,000 - 1,200		200 - 300 350 - 450 600 - 800	150 - 200 250 - 300 500 - 600	....
Cast iron							
Soft gray	HSS Cast-alloys Sintered carbides	.... 450 - 600	120 - 150 225 - 300 350 - 450	90 - 120 160 - 220 250 - 350	75 - 90 125 - 160 200 - 250	35 - 75 70 - 125 100 - 200
Medium and malleable	HSS Cast-alloys Sintered carbides	.... 350 - 450	120 - 150 100 - 225 250 - 350	90 - 120 150 - 190 200 - 250	60 - 90 120 - 150 150 - 200	30 - 60 60 - 120 75 - 150
Hard alloys	HSS Cast-alloys Sintered carbides	.... 250 - 300	90 - 125 120 - 170 150 - 250	60 - 90 80 - 120 100 - 150	40 - 60 55 - 80 75 - 100	20 - 40 35 - 55 50 - 75
Chilled	HSS Cast-alloys Sintered carbides	10 - 15 30 - 50	.... 10 - 30	....	....	....
Copper base alloys							
Leaded, free-cutting, soft brass and bronze	HSS Cast-alloys Sintered carbides	.... 1,000 - 1,250	300 - 400 500 - 600 800 - 1,000	225 - 300 400 - 500 650 - 800	150 - 255 325 - 400 500 - 650	100 - 150 200 - 325 300 - 500
Normal brass, bronze low alloy	HSS Cast-alloys Sintered carbides	.... 700 - 800	275 - 350 375 - 425 600 - 700	225 - 275 325 - 375 500 - 600	150 - 225 250 - 325 400 - 500	100 - 150 175 - 250 200 - 400
Tough copper, high tin & alum. bronzes, gilding.	HSS Cast-alloys Sintered carbides	.... 500 - 600	100 - 150 225 - 300 400 - 500	75 - 100 180 - 225 300 - 400	50 - 75 125 - 180 200 - 300	35 - 50 75 - 125 100 - 200
Light alloys							
Magnesium	HSS Cast-alloys Sintered carbides	500 - 750 700 - 1,000 1,250 - 2,000	350 - 500 500 - 700 800 - 1,250	275 - 350 400 - 500 600 - 800	200 - 275 300 - 400 500 - 600	125 - 200 200 - 300 300 - 500
Aluminum	HSS Cast-alloys Sintered carbides	350 - 500 450 - 650 700 - 1,000	225 - 350 300 - 450 450 - 700	150 - 225 225 - 300 300 - 450	100 - 150 150 - 225 200 - 300	50 - 100 75 - 150 100 - 200
Titanium							
Pure & low alloys	HSS Cast-alloys Sintered carbides	.... 550 - 900	100 - 160 165 - 375 375 - 600	70 - 110 110 - 250 250 - 400	50 - 75 75 - 165 165 - 265	....

FIGURE 5.5 *(Continued)* Cuts, feeds, and speeds table.

Alpha alloys	HSS	....	30 - 75	20 - 50	....
	Cast-alloys	....	75 - 110	50 - 75	....
	Sintered carbides	165 - 450	110 - 300	75 - 200	50 - 135
Beta alloys	HSS	....	30 - 40	20 - 25	....
	Cast-alloys	....	40 - 90	25 - 60	....
	Sintered carbides	125 - 225	90 - 150	60 - 100	40 - 70
Plastics	HSS	....	....	....	....
Thermoplastic, thermosetting	Cast-alloys	....	....	....	....
	Sintered carbides	650 - 1,000	400 - 650	250 - 400	150 - 250
Abrasives	HSS	....	....	....	....
Glass, hard rubber, green	Cast-alloys	....	....	....	....
ceramics, marble.	Sintered carbides	150 - 250	75 - 150	....	....

NOTE: It is possible that a combination of speeds, feeds and cuts may be so selected for a given application, that a higher horsepower may be required than is available at the lathe spindle. In all cases, especially where combinations of values are selected that have not been used previously on a given machine, the selected values should have their required horsepower calculated. See the subsection (Horsepower requirements) for calculating required horsepower when speed, feed and cut are given. Values of depth of cut are in inches; feeds are given in ipr (inches per revolution) and speed is given in sfpm (surface feet per minute).

(Tabular values are in sfpm).

FIGURE 5.5 *(Continued)* Cuts, feeds, and speeds table.

Select the heaviest depth of cut and feed rate that the machine can sustain, considering its horsepower or kilowatt rating, in conjunction with the required surface finish desired on the workpiece.

Relation of Speed to Feed. The following general rules apply to most turning and boring operations:

- If the tool shows a built-up edge, increase feed or increase speed.
- If the tool shows excessive cratering, reduce feed or reduce speed.
- If the tool shows excessive edge wear, increase feed or reduce speed.

Caution. The productivity settings from the machining calculators and any handbook speed and feed tables are suggestions and guides only. *A safety hazard may exist* if the user calculates or uses a table-selected machine setting *without also considering* the machine power and the condition, size, strength, and rigidity of the workpiece, machine, and cutting tools.

5.2 THREADING AND THREAD SYSTEMS

Thread-turning inserts are available in different styles or types for turning external and internal thread systems such as UN series, 60° metric, Whitworth (BSW), Acme, ISO, American buttress, etc. Figure 5.6 shows some of the typical thread-cutting inserts.

The defining dimensions and forms for various thread systems are shown in Fig. 5.7*a* to *k* with indications of their normal industrial uses. The dimensions in the figure are in U.S. customary and metric systems as indicated. In all parts of the figure, P = pitch, reciprocal of threads per inch (for U.S. customary) or millimeters (for metric).

Figure 5.7*a* defines the ISO thread system: M (metric) and UN (unified national). *Typical uses:* All branches of the mechanical industries. Figure 5.7*b* defines the UNJ thread system (controlled-root radii). *Typical uses:* Aerospace industries. Figure 5.7*c* defines the Whitworth system (BSW). *Typical uses:* Fittings and pipe couplings for water, sewer, and gas lines. Presently replaced by ISO system. Figure 5.7*d* defines the American buttress system, 7° face. *Typical uses:* Machine design. Figure 5.7*e* defines the NPT (American national pipe thread) system. *Typical uses:* Pipe threads, fittings, and couplings. Figure 5.7*f* defines the BSPT (British standard pipe thread) system. *Typical uses:* Pipe thread for water, gas, and steam lines. Figure 5.7*g* defines the Acme thread system, 29°. *Typical uses:* Mechanical industries for motion-transmission screws. Figure 5.7*h* defines the stub Acme thread system, 29°. *Typical uses:* Same as Acme, but used where normal Acme thread is too deep. Figure 5.7*i* defines the API 1:6 tapered-thread system. *Typical uses:* Petroleum industries. Figure 5.7*j* defines the TR DIN 103 thread system. *Typical uses:* Mechanical industries for motion-transmission screws. Figure 5.7*k* defines the RD DIN 405 (round) thread system. *Typical uses:* Pipe couplings and fittings in the fire-protection and food industries.

Threading Operations. Prior to cutting (turning) any particular thread, the following should be determined:

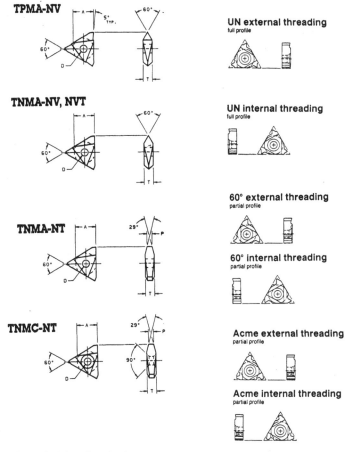

FIGURE 5.6 Typical thread cutting inserts.

- Machining toward the spindle (standard helix)
- Machining away from the spindle (reverse helix)
- Helix angle (see following equation)
- Insert and toolholder
- Insert grade
- Speed (sfpm)
- Number of thread passes
- Method of infeed

Calculating the Thread Helix Angle. To calculate the helix angle of a given thread system, use the following simple equation (see Fig. 5.8):

$$\tan \alpha = \frac{p}{\pi D_e}$$

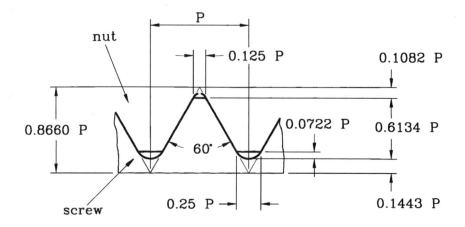

**(a) ISO - M (Metric)
(UN) (Unified National)**

FIGURE 5.7 Thread systems and dimensional geometry.

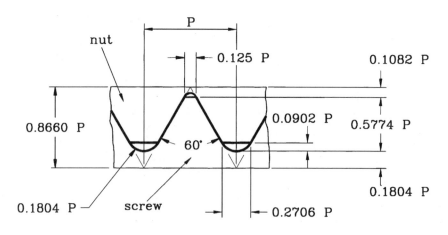

**(b) UNJ
(Controlled root radii)**

FIGURE 5.7 *(Continued)* Thread systems and dimensional geometry.

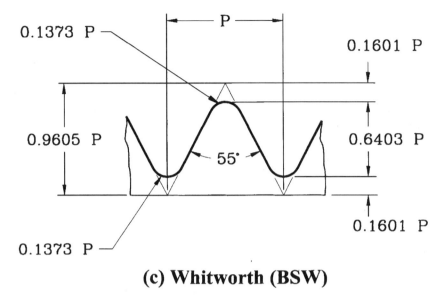

(c) Whitworth (BSW)

FIGURE 5.7 *(Continued)* Thread systems and dimensional geometry.

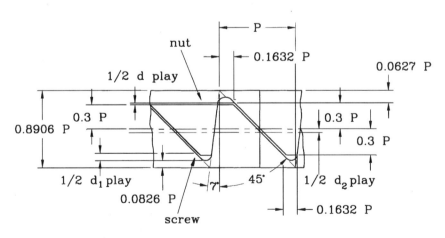

(d) American Buttress (7° face)

FIGURE 5.7 *(Continued)* Thread systems and dimensional geometry.

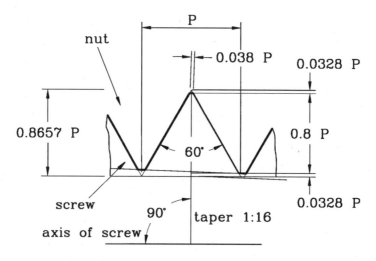

**(e) NPT
(American National Pipe Thread)**

FIGURE 5.7 *(Continued)* Thread systems and dimensional geometry.

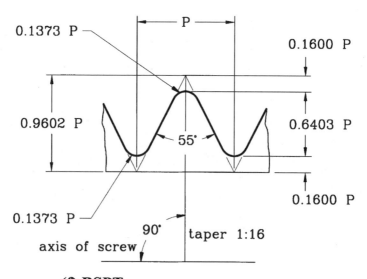

**(f) BSPT
(British Standard Pipe Thread)**

FIGURE 5.7 *(Continued)* Thread systems and dimensional geometry.

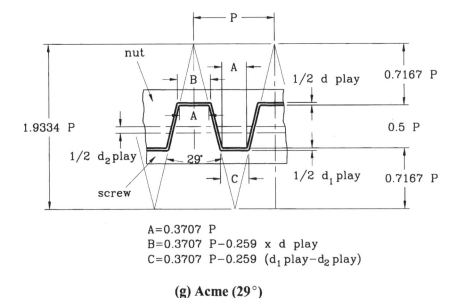

A=0.3707 P
B=0.3707 P−0.259 x d play
C=0.3707 P−0.259 (d₁ play−d₂ play)

(g) Acme (29°)

FIGURE 5.7 *(Continued)* Thread systems and dimensional geometry.

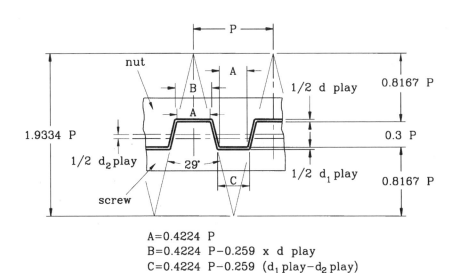

A=0.4224 P
B=0.4224 P−0.259 x d play
C=0.4224 P−0.259 (d₁ play−d₂ play)

(h) Acme Stub (29°)

FIGURE 5.7 *(Continued)* Thread systems and dimensional geometry.

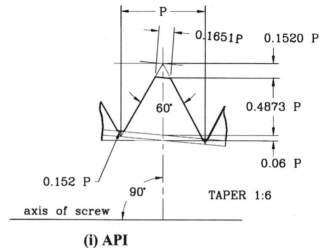

(i) API Taper: 1:6 (V-0.38"R)

FIGURE 5.7 *(Continued)* Thread systems and dimensional geometry.

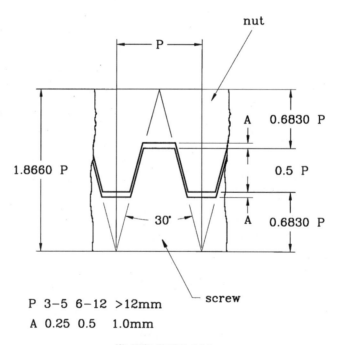

| P | 3–5 | 6–12 | >12mm |
| A | 0.25 | 0.5 | 1.0mm |

(j) TR DIN 103

FIGURE 5.7 *(Continued)* Thread systems and dimensional geometry.

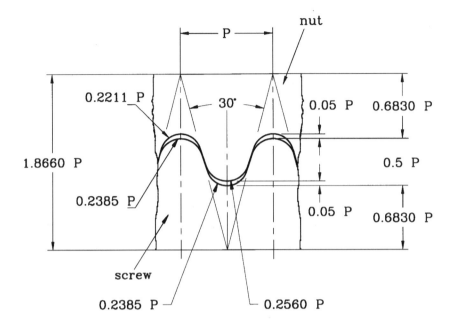

(k) RD DIN 405 (Round)

FIGURE 5.7 *(Continued)* Thread systems and dimensional geometry.

where tan α = natural tangent of the helix angle (natural function)
  D_e = effective diameter of thread, in or mm
  π = 3.1416
  p = pitch of thread, in or mm

EXAMPLE. Find the helix angle of a unified national coarse 0.375-16 thread, using the effective diameter of the thread:

$$p = \frac{1}{16} = 0.0625$$

(The pitch is the reciprocal of the number of threads per inch in the U.S. customary system.)

$$D_e = 0.375 \text{ in}$$

Therefore,

$$\tan \alpha = \frac{0.0625}{3.1416 \times 0.375} = \frac{0.0625}{1.1781} = 0.05305$$

$$\arctan 0.05305 = 3.037° \quad \text{or} \quad 3°2.22'$$

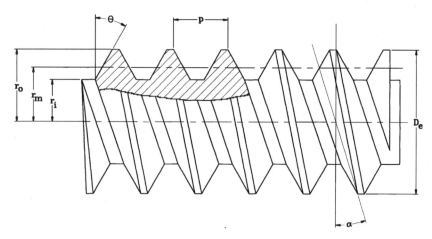

FIGURE 5.8 Calculating the helix angle α (alpha).

The helix angle of any helical thread system can be found by using the preceding procedure.

NOTE. For more data and calculations for threads, see Chap. 9.

Cutting Procedures for External and Internal Threads: Machine Setups. Figure 5.9 illustrates the methods for turning the external thread systems (standard and reverse helix). Figure 5.10 illustrates the methods for turning the internal thread systems (standard and reverse helix).

Problems in Thread Cutting

Problem	Possible remedy
Burr on crest of thread	1. Increase surface feet per minute (rpm).
	2. Use positive rake.
	3. Use full-profile insert (NTC type).
Poor tool life	1. Increase surface feet per minute (rpm).
	2. Increase chip load.
	3. Use more wear-resistant tool.
Built-up edge	1. Increase surface feet per minute (rpm).
	2. Increase chip load.
	3. Use positive rake, sharp tool.
	4. Use coolant or increase concentration.
Torn threads on workpiece	1. Use neutral rake.
	2. Alter infeed angle.
	3. Decrease chip load.
	4. Increase coolant concentration.
	5. Increase surface feet per minute (rpm).

FORMULAS AND CALCULATIONS FOR MACHINING OPERATIONS 5.21

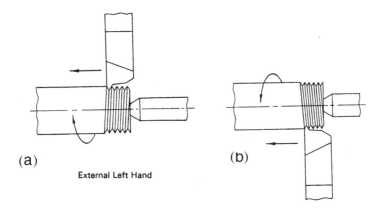

(a) External Left Hand

(b) External Right Hand

Feed Direction Towards Spindle (Standard Helix)

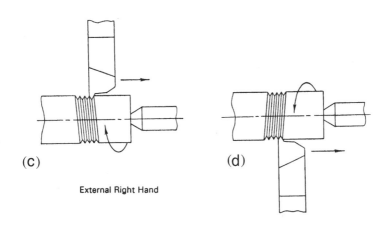

(c) External Right Hand

(d) External Left Hand

Feed Direction Away from Spindle (Reverse Helix)

FIGURE 5.9 Methods for cutting external threads.

5.22 CHAPTER FIVE

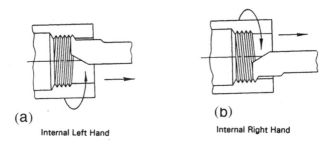

(a) Internal Left Hand

(b) Internal Right Hand

Feed Direction Away from Spindle (Reverse Helix)

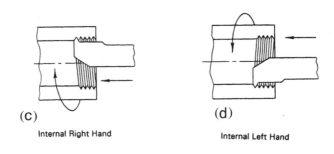

(c) Internal Right Hand

(d) Internal Left Hand

Feed Direction Towards Spindle (Standard Helix)

FIGURE 5.10 Methods for cutting internal threads.

5.3 MILLING

Milling is a machining process for generating machined surfaces by removing a predetermined amount of material progressively from the workpiece. The milling process employs relative motion between the workpiece and the rotating cutting tool to generate the required surfaces. In some applications the workpiece is stationary and the cutting tool moves, while in others the cutting tool and the workpiece are moved in relation to each other and to the machine. A characteristic feature of the milling process is that each tooth of the cutting tool takes a portion of the stock in the form of small, individual chips.

Typical cutting tool types for milling-machine operations are shown in Figs. 5.11*a* to *l*.

FORMULAS AND CALCULATIONS FOR MACHINING OPERATIONS 5.23

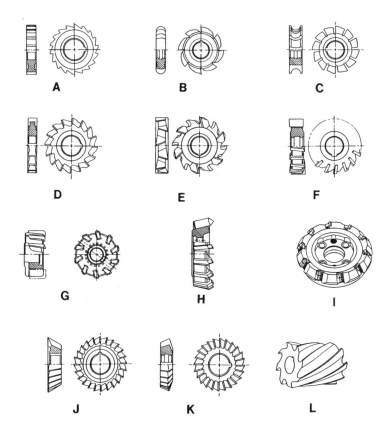

Milling Cutter Styles - High-Speed Steel and Carbide Insert

A Disk type milling cutter
B Convex half-round milling cutter
C Concave half-round milling cutter
D Three-side milling cutter
E Staggered-tooth milling cutter
F Inserted blade milling cutter
G Face milling cutter
H Face milling head
I Double-angle carbide insert milling cutter
J Single-angle milling cutter
K Double-angle milling cutter
L Left hand slab milling cutter

FIGURE 5.11 Typical cutting tools for milling.

Milling Methods

- Peripheral milling (slab milling)
- Face milling and straddle milling
- End milling
- Single-piece milling
- String or "gang" milling

- Slot milling
- Profile milling
- Thread milling
- Worm milling
- Gear milling

Modern milling machines have many forms, but the most common types are shown in Figs. 5.12 and 5.13. The well-known and highly popular Bridgeport-type milling machine is shown in Fig. 5.12. The Bridgeport machine is often used in tool and die making operations and in model shops, where prototype work is done. The great stability and accuracy of the Bridgeport makes this machine popular with

FIGURE 5.12 The Bridgeport milling machine.

FORMULAS AND CALCULATIONS FOR MACHINING OPERATIONS 5.25

FIGURE 5.13 Modern CNC machining center.

many experienced machinists and die makers. The Bridgeport shown in Fig. 5.12 is equipped with digital sensing controls and read-out panel, reading to ±0.0005 in.

The modern machining center is being used to replace the conventional milling machine in many industrial applications. Figure 5.13 shows a machining center, with its control panel at the right side of the machine. Machines such as these generally cost $250,000 or more depending on the accessories and auxiliary equipment obtained with the machine. These machines are the modern workhorses of industry and cannot remain idle for long periods owing to their cost.

The modern machining center may be equipped for three-, four-, or five-axis operation. The normal or common operations usually call for three-axis machining, while more involved machining procedures require four- or even five-axis operation. Three-axis operation consists of x and y table movements and z-axis vertical spindle movements. The four-axis operation includes the addition of spindle rotation with three-axis operation. Five-axis operation includes a horizontal fixture for rotating the workpiece on a horizontal axis at a predetermined speed (rpm), together with the functions of the four-axis machine. This allows all types of screw threads to be machined on the part and other operations such as producing a worm for worm-gear applications, segment cuts, arcs, etc. Very complex parts may be mass produced economically on a three-, four-, or five-axis machining center, all automatically, using computer numerical control (CNC).

The control panels on these machining centers contain a microprocessor that is, in turn, controlled by a host computer, generally located in the tool or manufacturing engineering office; the host computer controls one or more machines with direct numerical control (DNC) or distributed numerical control. Various machining programs are available for writing the operational instructions sent to the controller on the machining center. Figure 5.14 shows a detailed view of a typical microprocessor (CNC) control panel used on a machining center. This particular control panel is from an Enshu 550-V machining center, a photograph of which appears in Fig. 5.13.

Milling Calculations. The following calculation methods and procedures for milling operations are intended to be guidelines and not *absolute* because of the many variables encountered in actual practice.

Metal-Removal Rates. The *metal-removal rate* R (sometimes indicated as mrr) for all types of milling is equal to the volume of metal removed by the cutting process in a given time, usually expressed as cubic inches per minute (in^3/min). Thus,

$$R = WHf$$

where R = metal-removal rate, in^3/min.
  W = width of cut, in
  H = depth of cut, in
  f = feed rate, inches per minute (ipm)

FIGURE 5.14 The control panel from machine shown in Fig. 5.13.

FORMULAS AND CALCULATIONS FOR MACHINING OPERATIONS   5.27

In peripheral or slab milling, W is measured parallel to the cutter axis and H perpendicular to the axis. In face milling, W is measured perpendicular to the axis and H parallel to the axis.

Feed Rate. The speed or rate at which the workpiece moves past the cutter is the *feed rate* f, which is measured in inches per minute (ipm). Thus,

$$f = F_t N C_{rpm}$$

where f = feed rate, ipm
F_t = feed per tooth (chip thickness), in or cpt
N = number of cutter teeth
C_{rpm} = rotation of the cutter, rpm

Feed per Tooth. Production rates of milled parts are directly related to the feed rate that can be used. The feed rate should be as high as possible, considering machine rigidity and power available at the cutter. To prevent overloading the machine drive motor, the *feed per tooth allowable* F_t may be calculated from

$$F_t = \frac{K h p_c}{N C_{rpm} W H}$$

where hp_c = horsepower available at the cutter (80 to 90 percent of motor rating), i.e., if motor nameplate states 15 hp, then hp available at the cutter is 0.8 to 0.9×15 (80 to 90 percent represents motor efficiency)
K = machinability factor (see Fig. 5.15)

Other symbols are as in preceding equation.

Figure 5.16 gives the suggested feed per tooth for milling using high-speed-steel (HSS) cutters for the various cutter types. For carbide, cermets, and ceramic tools, see the figures in the cutting tool manufacturers' catalogs.

Material	K(in³/min/(hp_c))
Cold drawn steel, SAE 1112, 1120, 1315	1.0
Forged and alloy steel, SAE 3120, 1020, 2320, 2345, 150-300 BHN	0.63 - 0.87
Alloy steel, 300 - 400 BHN	0.5
Malleable iron and cold drawn steel, SAE 6140	0.9
Cast irons:	
Soft	1.5
Medium	0.8 - 1.0
Hard	0.6 - 0.8
Stainless steel, AISI 416, free-machining	1.1
Stainless steel, austenitic, AISI 303, free-machining	0.83
Stainless steel, austenitic, AISI 304	0.72
Tool steel	0.51
Bronze and brass:	
Soft	1.7 - 2.5
Medium	1.0 - 1.4
Hard	0.6 - 1.0
Aluminum and magnesium	2.5 - 4.0
Monel metal	0.55
Copper, annealed	0.84
Nickel	0.53
Titanium & alloys	0.75 - 1.1

NOTE: "K" values are in cubic inches per minute per cutter horsepower (in³/min/hp_c)
"K" values for carbide cutters are approx. 25% higher.

FIGURE 5.15 K factor table.

Suggested Feed per Tooth for Milling - High-Speed Steel Cutters (Tabulated Data in Inches)

Material	Face mills	Helical mills	Slot/side mills	End mills	Form-relieved cutters
Magnesium & alloys	0.022	0.018	0.013	0.011	0.007
Aluminum & Alloys	0.022	0.018	0.013	0.011	0.007
Free-cutting brasses & bronzes	0.022	0.018	0.013	0.011	0.007
Medium brasses & bronzes	0.014	0.011	0.008	0.007	0.004
Hard brasses & bronzes	0.009	0.007	0.006	0.005	0.003
Copper	0.012	0.010	0.007	0.006	0.004
Cast iron, soft (150-180 Bhn)	0.016	0.013	0.009	0.008	0.005
Cast iron, medium (180-220 Bhn)	0.013	0.010	0.007	0.007	0.004
Cast iron, hard (220-300 Bhn)	0.011	0.008	0.006	0.006	0.003
Malleable iron	0.012	0.010	0.007	0.006	0.004
Cast steel	0.012	0.010	0.007	0.006	0.004
Low-carbon steel, free-machining	0.012	0.010	0.007	0.006	0.004
Low-carbon steels	0.010	0.008	0.006	0.005	0.003
Medium-carbon steels	0.010	0.008	0.006	0.005	0.003
Alloy steel, ann'ld (180-220 Bhn)	0.008	0.007	0.005	0.004	0.003
Alloy steel, tough (220-300 Bhn)	0.006	0.005	0.004	0.003	0.002
Alloy steel, hard (300-400 Bhn)	0.004	0.003	0.003	0.002	0.002
Stainless steels, free-machining	0.010	0.008	0.006	0.005	0.003
Stainless steels	0.006	0.005	0.004	0.003	0.002
Monel metal	0.008	0.007	0.005	0.004	0.003
Titanium & alloys	0.008	0.007	0.005	0.004	0.003
Machinable plastics	0.013	0.010	0.008	0.007	0.004

NOTE: Tabular data in inches. For feed per tooth in millimeters, multiply tabular data by 25.4. For carbon-steel cutters, multiply tabular data by 0.50 or divide by 2. Source: Cincinnati Milicron, Inc.

FIGURE 5.16 Milling feed table, HSS.

Cutting Speed. The *cutting speed* of a milling cutter is the peripheral linear speed resulting from the rotation of the cutter. The cutting speed is expressed in feet per minute (fpm or ft/min) or surface feet per minute (sfpm or sfm) and is determined from

$$S = \frac{\pi D(\text{rpm})}{12}$$

where S = cutting speed, fpm or sfpm (sfpm is also termed spm)
 D = outside diameter of the cutter, in
 rpm = rotational speed of cutter, rpm

The required rotational speed of the cutter may be found from the following simple equation:

$$\text{rpm} = \frac{S}{(D/12)\pi} \quad \text{or} \quad \frac{S}{0.26D}$$

When it is necessary to increase the production rate, it is better to change the cutter material rather than to increase the cutting speed. Increasing the cutting speed alone may shorten the life of the cutter, since the cutter is usually being operated at its maximum speed for optimal productivity.

General Rules for Selection of the Cutting Speed
- Use lower cutting speeds for longer tool life.
- Take into account the Brinell hardness of the material.
- Use the lower range of recommended cutting speeds when starting a job.
- For a fine finish, use a lower feed rate in preference to a higher cutting speed.

FORMULAS AND CALCULATIONS FOR MACHINING OPERATIONS **5.29**

Number of Teeth: Cutter. The number of cutter teeth N required for a particular application may be found from the simple expression (not applicable to carbide or other high-speed cutters)

$$N = \frac{f}{F_t C_{rpm}}$$

where f = feed rate, ipm
F_t = feed per tooth (chip thickness), in
C_{rpm} = rotational speed of cutter, rpm
N = number of cutter teeth

An industry-recommended equation for calculating the number of cutter teeth required for a particular operation is

$$N = 19.5 \sqrt{R} - 5.8$$

where N = number of cutter teeth
R = radius of cutter, in

This simple equation is suitable for HSS cutters only and is not valid for carbide, cobalt cast alloy, or other high-speed cutting tool materials.

Figure 5.17 gives recommended cutting speed ranges (sfpm) for HSS cutters. Check the cutting tool manufacturers' catalogs for feeds, speeds, etc. for advanced cutting tool materials (i.e., carbide, cermet, ceramic, etc.).

Milling Horsepower. Ratios for metal removal per horsepower (cubic inches per minute per horsepower at the milling cutter) have been given for various materials (see Fig. 5.17). The general equation is

Milling Cutting Speeds for Various Materials
(sfpm) Surface feet per minute (High-speed steel tools only)

Material	High-speed steel tools	
	Rough	Finish
Cast iron	50 - 60	80 - 110
Semisteel	40 - 50	65 - 90
Malleable iron	80 - 100	110 - 130
Cast steel	45 - 60	70 - 90
Copper	100 - 150	150 - 200
Brass	200 - 300	200 - 300
Bronze	100 - 150	150 - 180
Aluminum	400 - 450	700 - 750
* Magnesium	600 - 800	1,000 - 1,500
SAE steels:		
1020 (coarse feed), low-carbon	60 - 80	60 - 80
1020 (fine feed), low-carbon	100 - 120	100 - 120
1035, medium-carbon	75 - 90	90 - 120
1330, alloy steel	90 - 110	90 - 110
1050, Med-high-carbon	60 - 80	100 - 125
2315, nickel steel	90 - 110	90 - 110
3150, nickel-chromium	50 - 60	70 - 90
4150, chrome-molybdenum	40 - 50	70 - 90
4340, nickel-chrome-molybdenum	40 - 50	60 - 70
Stainless steel	60 - 80	100 - 120
Titanium, hard alloy	80 - 100	110 - 130

NOTE: Tabular data ranges are in sfpm (surface feet per minute for HSS cutters only). For carbide cutters, increase sfpm by 25% (min.).
* A fire hazard is present when machining magnesium at high-speeds.

FIGURE 5.17 Milling cutting speeds, HSS.

$$K = \frac{\text{in}^3/\text{min}}{\text{hp}_c} = \frac{WHf}{\text{hp}_c}$$

where K = metal removal factor, in³/min/hp$_c$ (see Fig. 5.17)
 hp$_c$ = horsepower at the cutter
 W = width of cut, in
 H = depth of cut, in
 f = feed rate, ipm

The total horsepower required at the cutter may then be expressed as

$$\text{hp}_c = \frac{\text{in}^3/\text{min}}{K} \quad \text{or} \quad \frac{WHf}{K}$$

The K factor varies with type and hardness of material, and for the same material varies with the feed per tooth, increasing as the chip thickness increases. The K factor represents a particular rate of metal removal and not a general or average rate. For a quick approximation of total power requirements at the machine motor, see Fig. 5.18, which gives the maximum metal-removal rates for different horsepower-rated milling machines cutting different materials.

Milling-Machine Horsepower Ratings - for maximum metal removal rates (in³/min) for HSS (high-speed steel cutters)

Workpiece Material	Rated hp of Machine								
	3	5	7.5	10	15	20	25	30	40	50
	Max. Metal Removal (in³/min)								
Aluminum...............	2.7	5.5	8.7	12	18	27	37	48	69	91
Brass, soft................	2.4	4.7	7.5	10	16	24	32	41	60	79
Bronze, hard............	1.7	3.3	5.3	7.3	11	17	23	30	43	56
Bronze, very hard....	0.78	1.6	2.5	3.4	5.3	7.8	11	15	20	26
Cast iron, soft..........	1.6	3.2	5.2	7.1	11	16	22	28	41	54
Cast iron, hard.........	1	2	3.3	4.6	7	10	14	18	26	35
Cast iron, chilled......	0.78	1.6	2.5	3.4	5.3	7.8	10	13	19	26
Malleable iron..........	1	2.1	3.4	4.7	7.3	11	14	18	26	36
Steel, soft.................	1	2	3.3	4.6	7	10	14	18	26	35
Steel, medium...........	0.78	1.6	2.5	3.4	5.3	7.8	10	13	19	26
Steel, hard................	0.56	1.1	1.8	2.5	3.9	5.7	7.7	10	14	19

FIGURE 5.18 Milling machine horsepower ratings.

Typical Milling Problem and Calculations

Problem. We want to slot or side mill the maximum amount of material, in³/min, from an aluminum alloy part with a milling machine rated at 5 hp at the cutter. The milling cutter has 16 teeth, and has a tooth width of 0.750 in.

Use the following calculations as a guide for milling different materials.

Solution. Since production rates of milled parts are directly related to the feed rate allowed, the feed rate f should be as high as possible for a particular machine. Feed rate, ipm, is expressed as:

$$f = F_t N C_{\text{rpm}}$$

FORMULAS AND CALCULATIONS FOR MACHINING OPERATIONS   5.31

To prevent overloading the machine drive motor, the feed per tooth allowable F_t, also called chip thickness (cpt), may be calculated as:

$$F_t = \frac{K h p_c}{N C_{rpm} W H}$$

After selecting the machinability factor K from Fig. 5.15 (for aluminum it is 2.5 to 4, or an average of 3.25), calculate the depth of cut H when the cutter C_{rpm} is 200 and the feed per tooth F_t is selected from Fig. 5.16 (i.e., 0.013 for slot or side milling). Solve the preceding equation for H:

$$F_t = \frac{3.25 \times 5}{16 \times 200 \times 0.75 \times H}$$

$$0.013 = \frac{16.25}{2400 H}$$

$$0.013 \times 2400 H = 16.25$$

$$31.2 H = 16.25$$

$$H = 0.521 \text{ in depth of cut}$$

The feed rate f, in/min, is then found from:

$$f = F_t N C_{rpm}$$

$$f = 0.013 \times 16 \times 200$$

$$f = 41.6 \text{ linear in/min}$$

The maximum metal removal rate R is then calculated from:

$$R = W H f$$

where f = feed rate = 41.6 in/min (previously calculated)
W = 0.750 in (given width of the milling cutter)
H = 0.521 (previously calculated depth of cut)

Then,

$$R = 0.750 \times 0.521 \times 41.6$$

$$R = 16.26 \text{ in}^3/\text{min}$$

The K factor for aluminum was previously listed as an average 3.25 in^3/min/hp. We previously listed the horsepower at the cutter as 5 hp. Then,

$$3.25 \times 5 = 16.25 \text{ in}^3/\text{min}$$

which agrees with the previously calculated $R = 16.26$ in^3/min.

The diameter of the cutter can then be calculated from:

$$S = \frac{\pi D(\text{rpm})}{12}$$

Selecting S, sfpm, from Fig. 5.17 as 400 for aluminum, and solving the preceding equation for the cutter diameter D:

$$400 = \frac{3.1416 \times D \times 200}{12}$$

$$628D = 4800$$

$$D = 7.6 \text{ in dia.}$$

Now, let us select a cutter of 6-in diameter, and recalculate S:

$$S = \frac{3.1414 \times 6 \times 200}{12}$$

$$12S = 3770$$

$$S = 314.2 \text{ sfpm}$$

which is allowable for aluminum, using HSS cutters.

NOTE. The preceding calculations are for high-speed steel (HSS) cutters. For carbide, ceramic, cermet, and advanced cutting tool materials, the cutter speed rpm can generally be increased by 25 percent or more, keeping the same feed per tooth F_t, where the higher rpm will increase the feed rate f and give higher productivity. Also, the recommended cutting parameters or values for depth of cut, surface speed, rpm of the cutter, and other data for the advanced cutting tool inserts are given in the cutting tool manufacturers' catalogs. These catalogs also list the various types and shapes of inserts for different materials to be cut and types of machining applications such as turning, boring, and milling.

Modern Theory of Milling. The key characteristics of the milling process are

- Simultaneous motion of cutter rotation and feed movement of the workpiece
- Interrupted cut
- Production of tapered chips

It was common practice for many years in the industry to mill *against* the direction of feed. This was due to the type of tool materials then available (HSS) and the absence of antibacklash devices on the machines. This method became known as *conventional* or *up milling* and is illustrated in Fig. 5.19b. *Climb milling* or *down milling* is now the preferred method of milling with advanced cutting tool materials such as carbides, cermets, CBN, etc. Climb milling is illustrated in Fig. 5.19a. Here, the insert enters the cut with some chip load and proceeds to produce a chip that

FORMULAS AND CALCULATIONS FOR MACHINING OPERATIONS 5.33

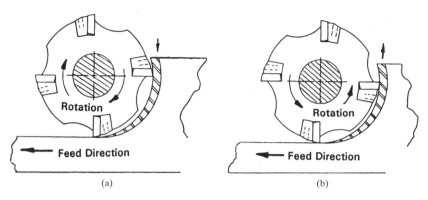

FIGURE 5.19 (*a*) Climb milling (preferred method); (*b*) up milling (conventional method).

thins as it progresses toward the end of the cut. This allows the heat generated in the cutting process to dissipate into the chip. Climb-milling forces push the workpiece toward the clamping fixture, in the direction of the feed. Conventional-milling (up-milling) forces are against the direction of feed and produce a lifting force on the workpiece and clamping fixture.

The angle of entry is determined by the position of the cutter centerline in relation to the edge of the workpiece. A negative angle of entry β is preferred and is illustrated in Fig. 5.20*b*, where the centerline of the cutter is below the edge of the workpiece. A negative angle is preferred because it ensures contact with the workpiece at the strongest point of the insert cutter. A positive angle of entry will increase insert chipping. If a positive angle of entry must be employed, use an insert with a honed or negative land.

Figure 5.20*a* shows an eight-tooth cutter climb milling a workpiece using a negative angle of entry, and the feed, or advance, per revolution is 0.048 in with a chip load per tooth of 0.006 in. The following milling formulas will allow you to calculate the various milling parameters.

In the following formulas,

nt = number of teeth or inserts in the cutter

cpt = chip load per tooth or insert, in

ipm = feed, inches per minute

fpr = feed (advance) per revolution, in

D = cutter effective cutting diameter, in

rpm = revolutions per minute

$sfpm$ = surface feet per minute (also termed sfm)

$$sfpm = \frac{\pi D (rpm)}{12} \qquad rpm = \frac{12(sfpm)}{\pi D} \qquad fpr = \frac{ipm}{rpm}$$

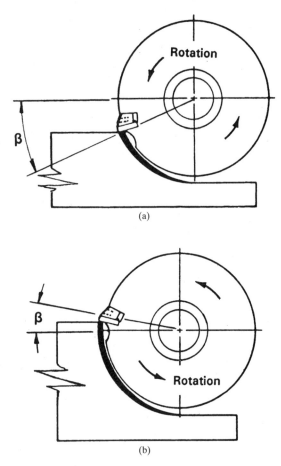

FIGURE 5.20 (*a*) Positive entry; (*b*) negative entry.

$$\text{ipm} = \text{cpt} \times \text{nt} \times \text{rpm} \qquad \text{cpt} = \frac{\text{ipm}}{\text{nt(rpm)}} \quad \text{or} \quad \frac{\text{fpr}}{\text{nt}}$$

EXAMPLE. Given a cutter of 5-in diameter, 8 teeth, 500 sfpm, and 0.007 cpt,

$$\text{rpm} = \frac{12 \times 500}{3.1416 \times 5} = 382$$

$$\text{ipm} = 0.007 \times 8 \times 382 = 21.4 \text{ in}$$

$$\text{fpr} = \frac{21.4}{382} = 0.056 \text{ in}$$

Slotting. Special consideration is given for slot milling, and the following equations may be used effectively to calculate chip load per tooth (cpt) and inches per minute (ipm):

$$\text{cpt} = \frac{[\sqrt{(D-x)}x/r](\text{ipm/rpm})}{\text{number of effective teeth}}$$

$$\text{ipm} = \text{rpm} \times \text{number of effective teeth} \left[\frac{\text{cpt}/\sqrt{(D-x)}x}{r} \right]$$

where D = diameter of slot cutter, in
r = radius of cutter, in
x = depth of slot, in
cpt = chip load per tooth, in
ipm = feed, inches per minute
rpm = rotational speed of cutter, rpms

Milling Horsepower for Advanced Cutting Tool Materials

Horsepower Consumption. It is advantageous to calculate the milling operational horsepower requirements before starting a job. Lower-horsepower machining centers take advantage of the ability of the modern cutting tools to cut at extremely high surface speeds (sfpm). Knowing your machine's speed and feed limits could be critical to your obtaining the desired productivity goals. The condition of your milling machine is also critical to obtaining these productivity goals. Older machines with low-spindle-speed capability should use the uncoated grades of carbide cutters and inserts.

Horsepower Calculation. A popular equation used in industry for calculating horsepower at the spindle is

$$\text{hp} = \frac{M_{rr} P_f}{E_s}$$

where M_{rr} = metal removal rate, in³/min
P_f = power constant factor (see Fig. 5.21b)
E_s = spindle efficiency, 0.80 to 0.90 (80 to 90 percent)

NOTE. The spindle efficiency is a reflection of losses from the machine's motor to actual power delivered at the cutter and must be taken into account, as the equation shows.

A table of P_f factors is shown in Fig. 5.21b.

NOTE. The metal removal rate M_{rr} = depth of cut × width of cut × ipm = in³/min.

Axial Cutting Forces at Various Lead Angles. Axial cutting forces vary as you change the lead angle of the cutting insert. The 0° lead angle produces the minimum axial force into the part. This is advantageous for weak fixtures and thin web sections. The 45° lead angle loads the spindle with the maximum axial force, which is advantageous when using the older machines.

Tangential Cutting Forces. The use of a tangential force equation is appropriate for finding the approximate forces that fixtures, part walls or webs, and the spindle bearings are subjected to during the milling operation. The tangential force is easily

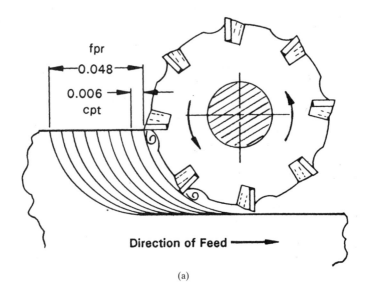

(a)

Power Constant Factor (P_f) for Milling Various Materials

Material	P_f Factor
Free-machining aluminum alloys	0.20
Gray cast irons	0.25
Non-ferrous free-machining alloys	0.45
Alloy cast irons and ductile irons	0.50
Martensitic stainless steels	0.81
Titanium	0.87
Free-machining carbon steel	0.89
Standard carbon steel	0.92
Alloy steels	0.95
Austenitic stainless steel	0.96
High-temperature alloys	1.00
Tool steels	1.05
Cobalt based alloys	1.20

Note: The P_f factors will vary per feed rate (ipm) and Brinell hardness (Bhn). The P_f factors in the table are for normal feed rates and material hardness ranges to 285 Bhn.

(b)

FIGURE 5.21 (*a*) Milling principle; (*b*) power constants for milling.

calculated when you have determined the horsepower being used at the spindle or cutter. It is important to remember that the tangential forces decrease as the spindle speed (rpm) increases, i.e., at higher surface feet per minute. The ability of the newer advanced cutting tools to operate at higher speeds thus produces fewer fixture- and web-deflecting forces with a decrease in horsepower requirements for any particular machine. Some of the new high-speed cutter inserts can operate efficiently at speeds of 10,000 sfpm or higher when machining such materials as free-machining aluminum and magnesium alloys.

The tangential force developed during the milling operations may be calculated from

FORMULAS AND CALCULATIONS FOR MACHINING OPERATIONS 5.37

$$t_f = \frac{126{,}000 \text{ hp}}{D(\text{rpm})}$$

where t_f = tangential force, lbf
hp = horsepower at the spindle or cutter
D = effective diameter of cutter, in
rpm = rotational speed, rpm

The preceding calculation procedure for finding the tangential forces developed on the workpiece being cut may be used in conjunction with the clamping fixture types and clamping calculations shown in Sec. 11.4, "Clamping Mechanisms and Calculation Procedures."

Cutter Speed, rpm, from Surface Speed, sfpm. A time-saving table of surface speed versus cutter speed is shown in Fig. 5.22 for cutter diameters from 0.25 through 5 in. For cutter speed rpm values when the surface speed is greater than 200 sfpm, use the simple equation

$$\text{rpm} = \frac{12(\text{sfpm})}{\pi D}$$

where D is the effective diameter of cutter in inches.

Applying Range of Conditions: Milling Operations. A convenient chart for modifying the speed and feed during a milling operation is shown in Fig. 5.23. As an example, if there seems to be a problem during a finishing cut on a milling operation, follow the arrows in the chart, and increase the speed while lowering the feed. For longer tool life, lower the speed while maintaining the same feed.

Diameter of cutter (in.)	Surface speed (ft. per min.)															
	25	30	35	40	50	55	60	70	75	80	90	100	120	140	160	180	200
	Cutter revolutions per minute															
1/4	382	458	535	611	764	851	917	1,070	1,147	1,222	1,376	1,528	1,834	2,139	2,445	2,750	3,056
5/16	306	367	428	489	611	672	733	856	917	978	1,100	1,222	1,466	1,711	1,955	2,200	2,444
3/8	255	306	357	408	509	560	611	713	764	815	916	1,018	1,222	1,425	1,629	1,832	2,036
7/16	218	262	306	349	437	481	524	611	656	699	786	874	1,049	1,224	1,398	1,573	1,748
1/2	191	229	268	306	382	420	459	535	573	611	688	764	917	1,070	1,222	1,375	1,528
5/8	153	184	214	245	306	337	367	428	459	489	552	612	736	857	979	1,102	1,224
3/4	127	153	178	203	254	279	306	357	381	408	458	508	610	711	813	914	1,016
7/8	109	131	153	175	219	241	262	306	329	349	392	438	526	613	701	788	876
1	95.5	115	134	153	191	210	229	267	287	306	344	382	458	535	611	688	764
1-1/4	76.3	91.8	107	123	153	168	183	214	230	245	274	306	367	428	490	551	612
1-1/2	63.7	76.3	89.2	102	127	140	153	178	191	204	230	254	305	356	406	457	508
1-3/4	54.5	65.5	76.4	87.3	109	120	131	153	164	175	196	218	262	305	349	392	436
2	47.8	57.3	66.9	76.4	95.5	105	115	134	143	153	172	191	229	267	306	344	382
2-1/2	38.2	45.8	53.5	61.2	76.3	84.2	91.7	107	114	122	138	153	184	213	245	275	306
3	31.8	38.2	44.6	51	63.7	69.9	76.4	89.1	95.3	102	114	127	152	178	208	228	254
3-1/2	27.3	32.7	38.2	44.6	54.5	60	65.5	76.4	81.8	87.4	98.1	109	131	153	174	196	218
4	23.9	28.7	33.4	38.2	47.8	52.6	57.3	66.9	71.7	76.4	86	95.6	115	134	153	172	191
5	19.1	22.9	26.7	30.6	38.2	42	45.9	53.5	57.3	61.1	68.8	76.4	91.7	107	122	138	153

NOTE: Tabular values are in revolutions per minute (rpm).

FIGURE 5.22   Cutter revolutions per minute from surface speed.

-GENERAL APPLICATIONS FOR CUTTING CONDITIONS-		
CONDITION	-SPEED-	-FEED-
Roughing	⇩	⇧
Finishing	⇧	⇩
End Milling	⇧	⇩
Slotting	⇧	⇩
Hard Material	⇩	⇨
Soft Material	⇧	⇧
Scale	⇩	⇧
Tool Life	⇩	⇨
Heavy d.o.c.	⇩	⇩

Higher- ⇧

Lower- ⇩

Same- ⇨

FIGURE 5.23 Applying range of conditions—milling operations.

5.4 DRILLING AND SPADE DRILLING

Drilling is a machining operation for producing round holes in metallic and nonmetallic materials. A *drill* is a rotary-end cutting tool with one or more cutting edges or lips and one or more straight or helical grooves or flutes for the passage of chips and cutting fluids and coolants. When the depth of the drilled hole reaches three or four times the drill diameter, a reduction must be made in the drilling feed and speed. A coolant-hole drill can produce drilled depths to eight or more times the diameter of the drill. The gundrill can produce an accurate hole to depths of more than 100 times the diameter of the drill with great precision.

Enlarging a drilled hole for a portion of its depth is called counterboring, while a counterbore for cleaning the surface a small amount around the hole is called *spotfacing*. Cutting an angular bevel at the perimeter of a drilled hole is termed *countersinking*. Countersinking tools are available to produce 82°, 90°, and 100° countersinks and other special angles.

FORMULAS AND CALCULATIONS FOR MACHINING OPERATIONS **5.39**

Drills are classified by material, length, shape, number, and type of helix or flute, shank, point characteristics, and size series. Most drills are made for right-hand rotation. Right-hand drills, as viewed from their point, with the shank facing away from your view, are rotated in a counterclockwise direction in order to cut. Left-hand drills cut when rotated clockwise in a similar manner.

Drill Types or Styles

- HSS jobber drills
- Solid-carbide jobber drills
- Carbide-tipped screw-machine drills
- HSS screw-machine drills
- Carbide-tipped glass drills
- HSS extralong straight-shank drills (24 in)
- Taper-shank drills (0 through number 7 ANSI taper)
- Core drills
- Coolant-hole drills
- HSS taper-shank extralong drills (24 in)
- Aircraft extension drills (6 and 12 in)
- Gun drills
- HSS half-round jobber drills
- Spotting and centering drills
- Parabolic drills
- S-point drills
- Square solid-carbide die drills
- Spade drills
- Miniature drills
- Microdrills and microtools

Drill Point Styles and Angles. Over a period of many years, the metalworking industry has developed many different drill point styles for a wide variety of applications from drilling soft plastics to drilling the hardest types of metal alloys. All the standard point styles and special points are shown in Fig. 5.24, including the important point angles which differentiate these different points. New drill styles are being introduced periodically, but the styles shown in Fig. 5.24 include some of the newer types as well as the commonly used older configurations.

The old practice of grinding drill points by hand and eye is, at the least, ineffective with today's modern drills and materials. For a drill to perform accurately and efficiently, modern drill-grinding machines such as the models produced by the Darex Corporation are required. Models are also produced which are also capable of sharpening taps, reamers, end mills, and countersinks.

Recommended general uses for drill point angles shown in Fig. 5.24 are shown here. Figure 5.24*k* illustrates web thinning of a standard twist drill.

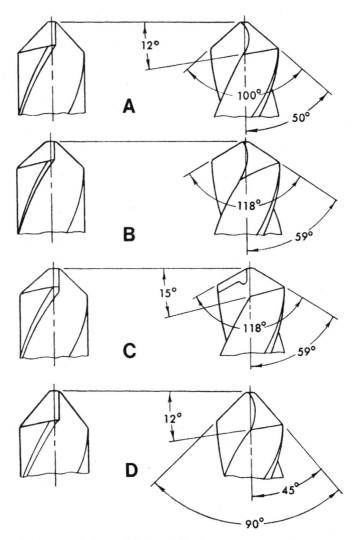

FIGURE 5.24 Drill-point styles and angles.

Typical Uses

A   Copper and medium to soft copper alloys
B   Molded plastics, Bakelite, etc.
C   Brasses and soft bronzes
D   Alternate for G, cast irons, die castings, and aluminum
E   Crankshafts and deep holes
F   Manganese steel and hard alloys (point angle 125 to 135°)

FORMULAS AND CALCULATIONS FOR MACHINING OPERATIONS 5.41

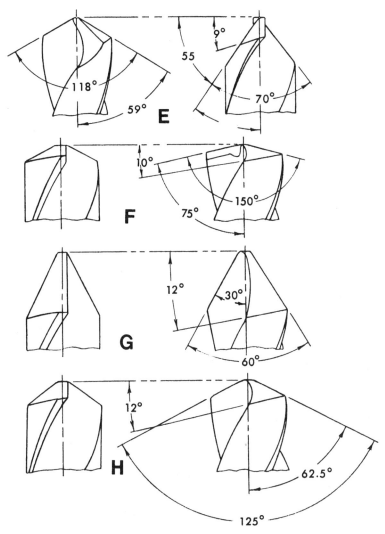

FIGURE 5.24 *(Continued)* Drill-point styles and angles.

G   Wood, fiber, hard rubber, and aluminum
H   Heat-treated steels and drop forgings
I   Split point, 118° or 135° point, self-centering (CNC applications)
J   Parabolic flute for accurate, deep holes and rapid cutting
K   Web thinning (thin the web as the drill wears from resharpening; this restores the chisel point to its proper length)

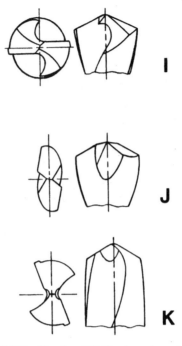

FIGURE 5.24 *(Continued)* Drill-point styles and angles.

Other drill styles which are used today include the helical or S-point, which is self-centering and permits higher feed rates, and the chamfered point, which is effective in reducing burr generation in many materials.

Drills are produced from high-speed steel (HSS) or solid carbide, or are made with carbide brazed inserts. Drill systems are made by many of the leading tool manufacturers which allow the use of removable inserts of carbide, cermet, ceramics, and cubic boron nitride (CBN). Many of the HSS twist drills used today have coatings such as titanium nitride, titanium carbide, aluminum oxide, and other tremendously hard and wear-resistant coatings. These coatings can increase drill life by as much as three to five times over premium HSS and plain-carbide drills.

Conversion of Surface Speed to Revolutions per Minute for Drills

Fractional Drill Sizes. Figure 5.25 shows the standard fractional drill sizes and the revolutions per minute of each fractional drill size for various surface speeds. The drilling speed tables that follow give the allowable drilling speed (sfpm) of the various materials. From these values, the correct rpm setting for drilling can be ascertained using the speed/rpm tables given here.

Wire Drill Sizes (1 through 80). See Fig. 5.26a and b.

Letter Drill Sizes. See Fig. 5.27.

FORMULAS AND CALCULATIONS FOR MACHINING OPERATIONS   5.43

FRACTIONAL SIZE DRILLS
Surface Feet per Minute

Diam. Inches	10'	12'	15'	20'	25'	30'	35'	40'	45'	50'	60'	70'	80'	90'	100'
							Revolutions per Minute								
1/64	2445	2934	3667	4889	6112	7334	8556	9778	11001	12223	14668	17112	19557	22001	24446
1/32	1222	1467	1833	2445	3056	3667	4278	4889	5500	6112	7334	8556	9778	11001	12223
3/64	815	978	1222	1630	2037	2445	2852	3259	3667	4074	4889	5704	6519	7334	8149
1/16	611	733	917	1222	1528	1833	2139	2445	2750	3056	3667	4278	4889	5500	6112
5/64	489	587	733	978	1222	1467	1711	1956	2200	2445	2934	3422	3911	4400	4889
3/32	407	489	611	815	1019	1222	1426	1630	1833	2037	2445	2852	3259	3667	4074
7/64	349	419	524	698	873	1048	1222	1397	1572	1746	2095	2445	2794	3143	3492
1/8	306	367	458	611	764	917	1070	1222	1375	1528	1833	2139	2445	2750	3056
9/64	272	326	407	543	679	815	951	1086	1222	1358	1630	1901	2173	2445	2716
5/32	244	293	367	489	611	733	856	978	1100	1222	1467	1711	1956	2200	2445
11/64	222	267	333	444	556	667	778	889	1000	1111	1333	1556	1778	2000	2222
3/16	204	244	306	407	509	611	713	815	917	1019	1222	1426	1630	1833	2037
13/64	188	226	282	376	470	564	658	752	846	940	1128	1316	1504	1692	1880
7/32	175	210	262	349	437	524	611	698	786	873	1048	1222	1397	1572	1746
15/64	163	196	244	326	407	489	570	652	733	815	978	1141	1304	1467	1630
1/4	153	183	229	306	382	458	535	611	688	764	917	1070	1222	1375	1528
9/32	136	163	204	272	340	407	475	543	611	679	815	951	1086	1222	1358
5/16	122	147	183	244	306	367	428	489	550	611	733	856	978	1100	1222
11/32	111	133	167	222	278	333	389	444	500	556	667	778	889	1000	1111
3/8	102	122	153	204	255	306	357	407	458	509	611	713	815	917	1019
13/32	94	113	141	188	235	282	329	376	423	470	564	658	752	846	940
7/16	87	105	131	175	218	262	306	349	393	437	524	611	698	786	873
15/32	81	98	122	163	204	244	285	326	367	407	489	570	652	733	815
1/2	76	92	115	153	191	229	267	306	344	382	458	535	611	688	764
9/16	68	81	102	136	170	204	238	272	306	340	407	475	543	611	679
5/8	61	73	92	122	153	183	214	244	275	306	367	428	489	550	611
11/16	56	67	83	111	139	167	194	222	250	278	333	389	444	500	556
3/4	51	61	76	102	127	153	178	204	229	255	306	357	407	458	509
13/16	47	56	71	94	118	141	165	188	212	235	282	329	376	423	470
7/8	44	52	65	87	109	131	153	175	196	218	262	306	349	393	437
15/16	41	49	61	81	102	122	143	163	183	204	244	285	326	367	407
1	38	46	57	76	95	115	134	153	172	191	229	267	306	344	382
1-1/8	34	41	51	68	85	102	119	136	153	170	204	238	272	306	340
1-1/4	31	37	46	61	76	92	107	122	138	153	183	214	244	275	306
1-3/8	28	33	42	56	69	83	97	111	125	139	167	194	222	250	278
1-1/2	25	31	38	51	64	76	89	102	115	127	153	178	204	229	255
1-5/8	24	28	35	47	59	71	82	94	106	118	141	165	188	212	235
1-3/4	22	26	33	44	55	65	76	87	98	109	131	153	175	196	218
1-7/8	20	24	31	41	51	61	71	81	92	102	122	143	163	183	204
2	19	23	29	38	48	57	67	76	86	95	115	134	153	172	191
2-1/4	17	20	25	34	42	51	59	68	76	85	102	119	136	153	170
2-1/2	15	18	23	31	38	46	53	61	69	76	92	107	122	138	153
2-3/4	14	17	21	28	35	42	49	56	63	69	83	97	111	125	139
3	13	15	19	25	32	38	45	51	57	64	76	89	102	115	127
3-1/2	11	13	16	22	27	33	38	44	49	55	65	76	87	98	109

For speeds higher than tabulated, multiply all values by 10 or 100. For speeds lower than tabulated, divide all values by 10.

FIGURE 5.25   Drill rpm/surface speed, fractional drills.

Tap-Drill Sizes for Producing Unified Inch and Metric Screw Threads and Pipe Threads

Tap-Drills for Unified Inch Screw Threads.   See Fig. 5.28.
Tap-Drill Sizes for Producing Metric Screw Threads.   See Fig. 5.29.
Tap-Drill Sizes for Pipe Threads (Taper and Straight Pipe).   See Fig. 5.30.

Equation for Obtaining Tap-Drill Sizes for Cutting Taps

$$D_h = D_{bm} - 0.0130 \left(\frac{\% \text{ of full thread desired}}{n_i} \right) \text{ for unified inch-size threads}$$

5.44 CHAPTER FIVE

WIRE SIZE DRILLS
Surface Feet per Minute

Diam. No.	10'	12'	15'	20'	25'	30'	35'	40'	45'	50'	60'	70'	80'	90'	100'
								Revolutions per Minute							
1	168	201	251	335	419	503	586	670	754	838	1005	1173	1340	1508	1675
2	173	207	259	346	432	519	605	691	778	864	1037	1210	1382	1555	1728
3	179	215	269	359	448	538	628	717	807	897	1076	1255	1434	1614	1793
4	183	219	274	366	457	548	640	731	822	914	1097	1280	1462	1645	1828
5	186	223	279	372	465	558	651	743	836	930	1115	1301	1487	1673	1859
6	187	225	281	374	468	562	655	749	843	936	1123	1310	1498	1685	1872
7	190	228	285	380	475	570	665	760	855	950	1140	1330	1520	1710	1900
8	192	230	288	384	480	576	672	768	864	960	1151	1343	1535	1727	1919
9	195	234	292	390	487	585	682	780	877	975	1169	1364	1559	1754	1949
10	197	237	296	395	494	592	691	790	888	987	1184	1382	1579	1777	1974
11	200	240	300	400	500	600	700	800	900	1000	1200	1400	1600	1800	2000
12	202	243	303	404	505	606	707	808	909	1010	1213	1415	1617	1819	2021
13	206	248	310	413	516	619	723	826	929	1032	1239	1450	1652	1859	2065
14	210	252	315	420	525	630	735	839	944	1050	1259	1469	1679	1889	2099
15	212	255	318	424	531	637	743	849	955	1064	1276	1489	1702	1914	2127
16	216	259	324	432	540	647	755	863	971	1079	1295	1511	1726	1942	2158
17	221	265	331	442	552	662	773	883	994	1104	1325	1546	1766	1987	2208
18	225	270	338	451	563	676	789	901	1014	1130	1356	1582	1808	2034	2260
19	230	276	345	460	575	690	805	920	1035	1151	1381	1611	1841	2071	2301
20	237	285	356	474	593	712	830	949	1068	1186	1423	1660	1898	2135	2372
21	240	288	360	480	601	721	841	961	1081	1201	1441	1681	1922	2162	2402
22	243	292	365	487	608	730	852	973	1095	1217	1460	1703	1946	2190	2433
23	248	298	372	496	620	744	868	992	1116	1240	1488	1736	1984	2232	2480
24	251	302	377	503	628	754	880	1005	1131	1257	1508	1759	2010	2262	2513
25	255	307	383	511	639	766	894	1022	1150	1276	1533	1789	2044	2300	2555
26	260	312	390	520	650	780	909	1039	1169	1299	1559	1819	2078	2338	2598
27	265	318	398	531	663	796	928	1061	1194	1327	1592	1857	2122	2388	2653
28	272	326	408	544	680	816	952	1087	1223	1360	1631	1903	2175	2447	2719
29	281	337	421	562	702	843	983	1123	1264	1405	1685	1966	2247	2528	2809
30	297	357	446	595	743	892	1040	1189	1338	1487	1784	2081	2378	2676	2973
31	318	382	477	637	796	955	1114	1273	1432	1592	1910	2228	2546	2865	3183
32	329	395	494	659	823	988	1152	1317	1482	1647	1976	2305	2634	2964	3293
33	338	406	507	676	845	1014	1183	1352	1521	1690	2028	2366	2704	3042	3380
34	344	413	516	688	860	1032	1204	1376	1549	1721	2065	2409	2753	3097	3442
35	347	417	521	694	868	1042	1215	1389	1563	1736	2083	2430	2778	3125	3472
36	359	430	538	717	897	1076	1255	1435	1614	1794	2152	2511	2870	3228	3587
37	367	441	551	735	918	1102	1285	1469	1653	1837	2204	2571	2938	3306	3673
38	376	452	564	753	941	1129	1317	1505	1693	1882	2258	2634	3010	3387	3763
39	384	461	576	768	960	1152	1344	1536	1728	1920	2303	2687	3071	3455	3839
40	390	468	585	780	974	1169	1364	1559	1754	1949	2339	2729	3118	3508	3898

For speeds higher than tabulated, multiply all values by 10 or 100. For speeds lower than tabulated, divide all values by 10.

FIGURE 5.26a Drill rpm/surface speed, wire-size drills.

$$D_{h1} = D_{bm1} - \left(\frac{\% \text{ of full thread desired}}{76.98}\right) \text{ for metric series threads}$$

where D_h = drilled hole size, in
 D_{h1} = drilled hole size, mm
 D_{bm} = basic major diameter of thread, in
 D_{bm1} = basic major diameter of thread, mm
 n_i = number of threads per inch

FORMULAS AND CALCULATIONS FOR MACHINING OPERATIONS 5.45

WIRE SIZE DRILLS
Surface Feet per Minute

Diam. No.	10'	12'	15'	20'	25'	30'	35'	40'	45'	50'	60'	70'	80'	90'	100'
							Revolutions per Minute								
41	398	477	597	796	995	1194	1393	1592	1790	1990	2387	2785	3183	3581	3979
42	409	490	613	817	1021	1226	1430	1634	1838	2043	2451	2860	3268	3677	4085
43	429	515	644	858	1073	1288	1502	1717	1931	2146	2575	3004	3434	3863	4292
44	444	533	666	888	1110	1332	1555	1777	1999	2221	2665	3109	3554	3999	4442
45	466	559	699	932	1165	1397	1630	1863	2096	2329	2795	3261	3726	4192	4658
46	472	566	707	943	1179	1415	1650	1886	2122	2358	2830	3301	3773	4244	4716
47	487	584	730	973	1216	1460	1703	1946	2190	2433	2920	3406	3893	4379	4866
48	503	603	754	1005	1256	1508	1759	2010	2262	2513	3016	3518	4021	4523	5026
49	523	628	785	1046	1308	1570	1831	2093	2355	2617	3140	3663	4186	4710	5233
50	546	655	819	1091	1364	1637	1910	2183	2456	2729	3274	3820	4366	4911	5457
51	570	684	855	1140	1425	1710	1995	2280	2565	2851	3421	3991	4561	5131	5701
52	602	722	902	1203	1504	1805	2105	2406	2707	3008	3609	4211	4812	5414	6015
53	642	770	963	1284	1605	1926	2247	2568	2889	3207	3848	4490	5131	5773	6414
54	694	833	1042	1389	1736	2083	2431	2778	3125	3473	4167	4862	5556	6251	6945
55	735	881	1102	1469	1836	2204	2571	2938	3306	3673	4408	5142	5877	6611	7346
56	821	986	1232	1643	2054	2464	2875	3286	3696	4108	4929	5751	6572	7394	8215
57	888	1066	1332	1777	2221	2665	3109	3553	3997	4452	5342	6232	7122	8013	8903
58	909	1091	1364	1819	2274	2728	3183	3638	4093	4547	5456	6367	7275	8186	9095
59	932	1118	1397	1863	2329	2795	3261	3726	4192	4658	5590	6521	7453	8388	9316
60	955	1146	1432	1910	2387	2865	3342	3820	4297	4775	5729	6684	7639	8594	9549
61	979	1175	1469	1959	2449	2938	3428	3918	4407	4897	5876	6856	7835	8815	9794
62	1005	1206	1508	2010	2513	3016	3518	4021	4523	5025	6030	7035	8040	9045	10050
63	1032	1239	1549	2065	2581	3097	3613	4129	4646	5160	6192	7224	8256	9288	10320
64	1061	1273	1592	2122	2653	3183	3714	4244	4775	5305	6366	7427	8488	9549	10610
65	1091	1310	1637	2183	2728	3274	3820	4365	4911	5455	6546	7637	8728	9819	10910
66	1157	1389	1736	2315	2894	3472	4051	4630	5207	5790	6948	8106	9264	10422	11580
67	1194	1432	1790	2387	2984	3581	4178	4775	5371	5970	7164	8358	9552	10746	11940
68	1232	1479	1848	2464	3080	3696	4313	4929	5545	6160	7392	8624	9856	11088	12320
69	1308	1570	1962	2616	3270	3924	4578	5232	5887	6530	7836	9142	10488	11754	13060
70	1364	1637	2046	2728	3410	4093	4775	5457	6139	6820	8184	9548	10912	12276	13640
71	1469	1763	2204	2938	3673	4407	5142	5876	6611	7365	8838	10311	11784	13257	14730
72	1528	1833	2272	3056	3820	4584	5348	6112	6875	7640	9168	10696	12224	13752	15280
73	1592	1910	2387	3183	3979	4775	5570	6366	7162	7960	9552	11144	12736	14328	15920
74	1698	2037	2546	3395	4244	5093	5942	6791	7639	8510	10212	11914	13616	15318	17020
75	1819	2183	2728	3638	4547	5457	6366	7276	8185	9095	10914	12733	14552	16371	18190
76	1910	2292	2865	3820	4775	5730	6684	7639	8594	9550	11460	13370	15280	17190	19100
77	2122	2546	3183	4244	5305	6366	7427	8488	9549	10610	12732	14854	16976	19098	21220
78	2387	2865	3581	4775	5968	7162	8356	9549	10743	11935	14322	16709	19096	21483	23870
79	2634	3161	3951	5269	6586	7903	9220	10537	11854	13170	15804	18438	21072	23706	26340
80	2829	3395	4244	5659	7074	8488	9903	11318	12732	14150	16980	19810	22640	25470	28300

For speeds higher than tabulated, multiply all values by 10 or 100. For speeds lower than tabulated, divide all values by 10.

FIGURE 5.26b Drill rpm/surface speed, wire-size drills.

NOTE. In the preceding equations, use the percentage whole number; i.e., for 84 percent, use 84.

EXAMPLE. What is the drilled hole size in inches for a ⅜-16 tapped thread with 84 percent of full thread?

$$D_h = 0.375 - 0.0130 \times \frac{84}{16} = 0.375 - 0.06825 = 0.30675 \text{ in}$$

5.46 CHAPTER FIVE

LETTER SIZE DRILLS
Surface Feet per Minute

Letter Size	10'	12'	15'	20'	25'	30'	35'	40'	45'	50'	60'	70'	80'	90'	100'
								Revolutions per Minute							
A	163	196	245	326	408	490	571	653	735	818	982	1145	1309	1472	1636
B	160	193	241	321	401	481	562	642	722	803	963	1124	1284	1445	1605
C	158	189	237	316	395	473	552	631	710	789	947	1105	1262	1420	1578
D	155	186	233	311	388	466	543	621	699	778	934	1089	1245	1400	1556
E	153	183	229	306	382	458	535	611	687	764	917	1070	1222	1375	1528
F	149	178	223	297	372	446	520	595	669	743	892	1040	1189	1337	1486
G	146	176	220	293	366	439	512	585	659	732	878	1024	1170	1317	1463
H	144	172	215	287	359	431	503	574	646	718	862	1005	1149	1294	1436
I	140	169	211	281	351	421	492	562	632	702	842	983	1123	1264	1404
J	138	165	207	276	345	414	483	552	621	690	827	965	1103	1241	1379
K	136	163	204	272	340	408	476	544	612	680	815	951	1087	1223	1359
L	132	158	198	263	329	395	461	527	593	659	790	922	1054	1185	1317
M	129	155	194	259	324	388	453	518	583	648	777	907	1036	1166	1295
N	126	152	190	253	316	379	442	506	569	633	759	886	1012	1139	1265
O	121	145	181	242	302	363	423	484	544	605	725	846	967	1088	1209
P	118	142	177	237	296	355	414	473	532	592	710	828	946	1065	1183
Q	115	138	173	230	288	345	403	460	518	575	690	805	920	1035	1150
R	113	135	169	225	282	338	394	451	507	564	676	789	902	1014	1127
S	110	132	165	220	274	329	384	439	494	549	659	769	878	988	1098
T	107	128	160	213	267	320	373	427	480	533	640	746	853	959	1066
U	104	125	156	208	259	311	363	415	467	519	623	727	830	934	1038
V	101	122	152	203	253	304	355	405	456	507	608	709	810	912	1013
W	99	119	148	198	247	297	346	396	445	495	594	693	792	891	989
X	96	115	144	192	240	289	337	385	433	481	576	672	769	865	962
Y	95	113	142	189	236	284	331	378	425	473	567	662	756	851	945
Z	92	111	139	185	231	277	324	370	416	462	555	647	740	832	925

For speeds higher than tabulated, multiply all values by 10 or 100. For speeds lower than tabulated divide all values by 10.

FIGURE 5.27 Drill rpm/surface speed, letter-size drills.

0.30675 in is then the decimal equivalent of the required tap drill for 84 percent of full thread. Use the next closest drill size, which would be letter size N (0.302 in). The diameters of the American standard wire and letter-size drills are shown in Fig. 5.31. For metric drill sizes see Fig. 5.32.

When producing the tapped hole, be sure that the correct class of fit is satisfied, i.e., class 2B, 3B, interference fit, etc. The different classes of fits for the thread systems are shown in the section of standards of the American National Standards Institute (ANSI) and the American Society of Mechanical Engineers (ASME).

Speeds and Feeds, Drill Geometry, and Cutting Recommendations for Drills. The composite drilling table shown in Fig. 5.33 has been derived from data originated by the Society of Manufacturing Engineers (SME) and various major drill manufacturers.

Spade Drills and Drilling. Spade drills are used to produce holes ranging from 1 in to over 6 in in diameter. Very deep holes can be produced with spade drills, including core drilling, counterboring, and bottoming to a flat or other shape. The spade drill consists of the spade drill bit and holder. The holder may contain coolant holes through which coolant can be delivered to the cutting edges, under pressure, which cools the spade and flushes the chips from the drilled hole.

The standard point angle on a spade drill is 130°. The rake angle ranges from 10 to 12° for average-hardness materials. The rake angle should be 5 to 7° for hard

FORMULAS AND CALCULATIONS FOR MACHINING OPERATIONS   5.47

Tap size	Tap drill size	Decimal equiv. of tap drill, in	Theoretical percent of thread, %	Probable mean over-size, in	Probable hole size, in	Probable percent of thread, %
0–80	56	0.0465	83	0.0015	0.0480	74
	3⁄64	0.0469	81	0.0015	0.0484	71
	1.20 mm	0.0472	79	0.0015	0.0487	69
	1.25 mm	0.0492	67	0.0015	0.0507	57
1–64	54	0.0550	89	0.0015	0.0565	81
	1.45 mm	0.0571	78	0.0015	0.0586	71
	53	0.0595	67	0.0015	0.0610	59
1–72	1.5 mm	0.0591	77	0.0015	0.0606	68
	53	0.0595	75	0.0015	0.0610	67
	1.55 mm	0610	67	0.0015	0.0606	68
2–56	51	0.0670	82	0.0017	0.0687	74
	1.75 mm	0.0689	73	0.0017	0.0706	66
	50	0.0700	69	0.0017	0.0717	62
	1.80 mm	0.0709	65	0.0017	0.0726	58
2–64	50	0.0700	79	0.0017	0.0717	70
	1.80 mm	0.0709	74	0.0017	0.0726	66
	49	0.0730	64	0.0017	0.0747	56
3–48	48	0.0760	85	0.0019	0.0779	78
	5⁄64	0.0781	77	0.0019	0.0800	70
	47	0.0785	76	0.0019	0.0804	69
	2.00 mm	0.0787	75	0.0019	0.0806	68
	46	0.0810	67	0.0019	0.0829	60
	45	0.0820	63	0.0019	0.0839	56
3–56	46	0.0810	78	0.0019	0.0829	69
	45	0.0820	73	0.0019	0.0839	65
	2.10 mm	0.0827	70	0.0019	0.0846	62
	2.15 mm	0.0846	62	0.0019	0.0865	54
4–40	44	0.0860	80	0.0020	0.0880	74
	2.20 mm	0.0866	78	0.0020	0.0886	72
	43	0.0890	71	0.0020	0.0910	65
	2.30 mm	0.0906	66	0.0020	0.0926	60
4–48	2.35 mm	0.0925	72	0.0020	0.0926	72
	42	0.0935	68	0.0020	0.0955	61
	3⁄32	0.0938	68	0.0020	0.0958	60
	2.40 mm	0.0945	65	0.0020	0.0965	57
5–40	40	0.0980	83	0.0023	0.1003	76
	39	0.0995	79	0.0023	0.1018	71
	38	0.1015	72	0.0023	0.1038	65
	2.60 mm	0.1024	70	0.0023	0.1047	63

FIGURE 5.28 Tap-drill sizes, unified inch screw threads.

steels and 15 to 20° for soft, ductile materials. The back-taper angle should be 0.001 to 0.002 in per inch of blade depth. The outside diameter clearance angle is generally between 7 to 10°.

The cutting speeds for spade drills are normally 10 to 15 percent lower than those for standard twist drills. See the tables of drill speeds and feeds in the preceding section for approximate starting speeds. Heavy feed rates should be used with spade

Tap size	Tap drill size	Decimal equiv. of tap drill, in	Theoretical percent of thread, %	Probable mean over-size, in	Probable hole size, in	Probable percent of thread, %
5–44	38	0.1015	79	0.0023	0.1038	72
	2.60 mm	0.1024	77	0.0023	0.1047	69
	37	0.1040	71	0.0023	0.1063	63
6–32	37	0.1040	84	0.0023	0.1063	78
	36	0.1065	78	0.0023	0.1088	72
	7/64	0.1095	70	0.0026	0.1120	64
	35	0.1100	69	0.0026	0.1126	63
	34	0.1100	67	0.0026	0.1136	60
6–40	34	0.1110	83	0.0026	0.1136	75
	33	0.1130	77	0.0026	0.1156	69
	2.90 mm	0.1142	73	0.0026	0.1168	65
	32	0.1160	68	0.0026	0.1186	60
8–32	3.40 mm	0.1339	74	0.0029	0.1368	67
	29	0.1360	69	0.0029	0.1389	62
8–36	29	0.1360	78	0.0029	0.1389	70
	3.5 mm	0.1378	72	0.0029	0.1407	65
10–24	27	0.1440	85	0.0032	0.1472	79
	3.70 mm	0.1457	82	0.0032	0.1489	76
	26	0.1470	79	0.0032	0.1502	74
	25	0.1495	75	0.0032	0.1527	69
	24	0.1520	70	0.0032	0.1552	64
10–32	5/32	0.1563	83	0.0032	0.1595	75
	22	0.1570	81	0.0032	0.1602	73
	21	0.1590	76	0.0032	0.1622	68
12–24	11/64	0.1719	82	0.0035	0.1754	75
	17	0.1730	79	0.0035	0.1765	73
	16	0.1770	72	0.0035	0.1805	66
12–28	16	0.1770	84	0.0035	0.1805	77
	15	0.1800	78	0.0035	0.1835	70
	4.60 mm	0.1811	75	0.0035	0.1846	67
	14	0.1820	73	0.0035	0.1855	66
1/4–20	9	0.1960	83	0.0038	0.1998	77
	8	0.1990	79	0.0038	0.2028	73
	7	0.2010	75	0.0038	0.2048	70
	13/64	0.2031	72	0.0038	0.2069	66
1/4–28	5.40 mm	0.2126	81	0.0038	0.2164	72
	3	0.2130	80	0.0038	0.2168	72
5/16–18	F	0.2570	77	0.0038	0.2608	72
	6.60 mm	0.2598	73	0.0038	0.2636	68
	G	0.2610	71	0.0041	0.2651	66
5/16–24	H	0.2660	86	0.0041	0.2701	78
	6.80 mm	0.2677	83	0.0041	0.2718	75

FIGURE 5.28 *(Continued)* Tap-drill sizes, unified inch screw threads.

FORMULAS AND CALCULATIONS FOR MACHINING OPERATIONS   5.49

Tap size	Tap drill size	Decimal equiv. of tap drill, in	Theoretical percent of thread, %	Probable mean over-size, in	Probable hole size, in	Probable percent of thread, %
	I	0.2720	75	0.0041	0.2761	67
⅜–16	7.80 mm	0.3071	84	0.0044	0.3115	78
	7.90 mm	0.3110	79	0.0044	0.3154	73
	⁵⁄₁₆	0.3125	77	0.0044	0.3169	72
	O	0.3160	73	0.0044	0.3204	68
8–24	²¹⁄₆₄	0.3281	87	0.0044	0.3325	79
	8.40 mm	0.3307	82	0.0044	0.3351	74
	Q	0.3320	79	0.0044	0.3364	71
	8.50 mm	0.3346	75	0.0044	0.3390	67
⁷⁄₁₆–14	T	0.3580	86	0.0046	0.3626	81
	²³⁄₆₄	0.3594	84	0.0046	0.3640	79
	9.20 mm	0.3622	81	0.0046	0.3668	76
	9.30 mm	0.3661	77	0.0046	0.3707	72
	U	0.3680	75	0.0046	0.3726	70
	9.40 mm	0.3701	73	0.0046	0.3747	68
⁷⁄₁₆–20	W	0.3860	79	0.0046	0.3906	72
	²⁵⁄₆₄	0.3906	72	0.0046	0.3952	65
½–13	10.50 mm	0.4134	87	0.0047	0.4181	82
	²⁷⁄₆₄	0.4219	78	0.0047	0.4266	73
½–20	²⁹⁄₆₄	0.4531	72	0.0047	0.4578	65
⁹⁄₁₆–12	¹⁵⁄₃₂	0.4688	87	0.0048	0.4736	82
	³¹⁄₆₄	0.4844	72	0.0048	0.4892	68
⁹⁄₁₆–18	½	0.5000	87	0.0048	0.5048	80
⅝–11	¹⁷⁄₃₂	0.5313	79	0.0049	0.5362	75
⅝–18	⁹⁄₁₆	0.5625	87	0.0049	0.5674	80
¾–10	⁴¹⁄₆₄	0.6406	84	0.0050	0.6456	80
	²¹⁄₃₂	0.6563	72	0.0050	0.6613	68
¼–16	¹¹⁄₁₆	0.6875	77	0.0050	0.6925	71
	17.50 mm	0.6890	75	0.0050	0.6940	69
⅞–9	⁴⁹⁄₆₄	0.7656	76	0.0052	0.7708	72
⅞–14	⁵¹⁄₆₄	0.7969	84	0.0052	0.8021	79
1–8	⁵⁵⁄₆₄	0.8594	87	0.0059	0.8653	83
	⅞	0.8750	77	0.0059	0.8809	73
1–12	²⁹⁄₃₂	0.9063	87	0.0059	0.9122	81
	⁵⁹⁄₆₄	0.9219	72	0.0060	0.9279	67
1–14	⁵⁹⁄₆₄	0.9219	84	0.0060	0.9279	78
1⅛–7	³¹⁄₃₂	0.9688	84	0.0062	0.9750	81
	⁶³⁄₆₄	0.9844	76	0.0067	0.9911	72
1⅛–12	1¹⁄₃₂	1.0313	87	0.0071	1.0384	80

FIGURE 5.28 *(Continued)* Tap-drill sizes, unified inch screw threads.

Metric Tap size	Tap drill size	Decimal equiv. of tap drill, in	Theoretical percent of thread, %	Probable mean over-size, in	Probable hole size, in	Probable percent of thread, %
M1.6 × 0.35	1.20 mm	0.0472	88	0.0014	0.0486	80
	1.25 mm	0.0492	77	0.0014	0.0506	69
M2 × 0.4	1/16	0.0625	79	0.0015	0.0640	72
	1.60 mm	0.0630	77	0.0017	0.0647	69
	52	0.0635	74	0.0017	0.0652	66
M2.5 × 0.45	2.05 mm	0.0807	77	0.0019	0.0826	69
	46	0.0810	76	0.0019	0.0829	67
	45	0.0820	71	0.0019	0.0839	63
M3 × 0.5	40	0.0980	79	0.0023	0.1003	70
	2.5 mm	0.0984	77	0.0023	0.1007	68
	39	0.0995	73	0.0023	0.1018	64
M3.5 × 0.6	33	0.1130	81	0.0026	0.1156	72
	2.9 mm	0.1142	77	0.0026	0.1163	68
	32	0.1160	71	0.0026	0.1186	63
M4 × 0.7	3.2 mm	0.1260	88	0.0029	0.1289	80
	30	0.1285	81	0.0029	0.1314	73
	3.3 mm	0.1299	77	0.0029	0.1328	69
M4.5 × 0.75	3.7 mm	0.1457	82	0.0032	0.1489	74
	26	0.1470	79	0.0032	0.1502	70
	25	0.1495	72	0.0032	0.1527	64
M5 × 0.8	4.2 mm	0.1654	77	0.0032	0.1686	69
	19	0.1660	75	0.0032	0.1692	68
M6 × 1	10	0.1935	84	0.0038	0.1973	76
	9	0.1960	79	0.0038	0.1998	71
	5 mm	0.1968	77	0.0038	0.2006	70
	8	0.1990	73	0.0038	0.2028	65
M7 × 1	A	0.2340	81	0.0038	0.2378	74
	6 mm	0.2362	77	0.0038	0.2400	70
	B	0.2380	74	0.0038	0.2418	66
M8 × 1.25	6.7 mm	0.2638	80	0.0041	0.2679	74
	17/64	0.2656	77	0.0041	0.2697	71
	H	0.2660	77	0.0041	0.2701	70
	6.8 mm	0.2677	74	0.0041	0.2718	68
M10 × 1.5	8.4 mm	0.3307	82	0.0044	0.3351	76
	Q	0.3320	80	0.0044	0.3364	75
	8.5 mm	0.3346	77	0.0044	0.3390	71
M12 × 1.75	10.25 mm	0.4035	77	0.0047	0.4082	72
	Y	0.4040	76	0.0047	0.4087	71
	13/32	0.4062	74	0.0047	0.4109	69
M14 × 2	15/32	0.4688	81	0.0048	0.4736	76
	12 mm	0.4724	77	0.0048	0.4772	72

FIGURE 5.29 Tap-drill sizes, metric screw threads.

drilling. The table shown in Fig. 5.34 gives recommended feed rates for spade drilling various materials.

Horsepower and Thrust Forces for Spade Drilling. The following simplified equations will allow you to calculate the approximate horsepower requirements and thrust needed to spade drill various materials with different diameter spade drills. In

FORMULAS AND CALCULATIONS FOR MACHINING OPERATIONS   5.51

Metric Tap size	Tap drill size	Decimal equiv. of tap drill, in	Theoretical percent of thread, %	Probable mean over- size, in	Probable hole size, in	Probable percent of thread, %
M16 × 2	35/64	0.5469	81	0.0049	0.5518	76
	14 mm	0.5512	77	0.0049	0.5561	72
M20 × 2.5	11/16	0.6875	78	0.0050	0.6925	74
	17.5 mm	0.6890	77	0.0052	0.6942	73
M24 × 3	13/16	0.8125	86	0.0052	0.8177	82
	21 mm	0.8268	76	0.0054	0.8322	73
	53/64	0.8281	76	0.0054	0.8335	73
M30 × 3.5	1 1/32	1.0312	83	0.0071	1.0383	80
	25.1 mm	1.0394	79	0.0071	1.0465	75
	1 3/64	1.0469	75	0.0072	1.0541	70
M36 × 4	1 17/64	1.2656	74	Reaming recommended		

FIGURE 5.29 *(Continued)* Tap-drill sizes, metric screw threads.

order to do this, you must find the feed rate for your particular spade drill diameter, as shown in Fig. 5.34, and then select the P factor for your material, as tabulated in Fig. 5.35.

The following equations may then be used to estimate the required horsepower at the machine's motor and the thrust required in pounds force for the drilling process.

$$C_{hp} = P\left(\frac{\pi D^2}{4}\right)FN$$

Taper pipe		Straight pipe	
Thread	Drill	Thread	Drill
1/8–27	R	1/8–27	S
1/4–18	7/16	1/4–18	29/64
3/8–18	37/64	3/8–18	19/32
1/2–14	23/32	1/2–14	47/64
3/4–14	59/64	3/4–14	15/16
1–11 1/2	1 5/32	1–11 1/2	1 3/16
1 1/4–11 1/2	1 1/2	1 1/4–11 1/2	1 33/64
1 1/2–11 1/2	1 47/64	1 1/2–11 1/2	1 3/4
2–11 1/2	2 7/32	2–11 1/2	2 7/32
2 1/2–8	2 5/8	2 1/2–8	2 21/32
3–8	3 1/4	3–8	3 9/32
3 1/2–8	3 3/4	3 1/2–8	3 25/32
4–8	4 1/4	4–8	4 9/32

FIGURE 5.30 Pipe taps.

DRILL NO.	DECIMAL	DRILL NO.	DECIMAL	DRILL NO.	DECIMAL
97	0.0059	56	0.0465	15	0.180
96	0.0063	55	0.052	14	0.182
95	0.0067	54	0.055	13	0.185
94	0.0071	53	0.0595	12	0.189
93	0.0075	52	0.0635	11	0.191
92	0.0079	51	0.067	10	0.1935
91	0.0083	50	0.070	9	0.196
90	0.0087	49	0.073	8	0.199
89	0.0091	48	0.076	7	0.201
88	0.0095	47	0.0785	6	0.204
87	0.010	46	0.076	5	0.2055
86	0.0105	45	0.082	4	0.209
85	0.011	44	0.086	3	0.213
84	0.0115	43	0.089	2	0.221
83	0.012	42	0.0935	1	0.228
82	0.0125	41	0.096	A	0.234
81	0.013	40	0.098	B	0.238
80	0.0135	39	0.0995	C	0.242
79	0.0145	38	0.1015	D	0.246
78	0.016	37	0.104	E	0.250
77	0.018	36	0.1065	F	0.257
76	0.020	35	0.110	G	0.261
75	0.021	34	0.111	H	0.266
74	0.0225	33	0.113	I	0.272
73	0.024	32	0.116	J	0.277
72	0.025	31	0.120	K	0.281
71	0.026	30	0.1285	L	0.290
70	0.028	29	0.136	M	0.295
69	0.0292	28	0.1405	N	0.302
68	0.031	27	0.144	O	0.316
67	0.032	26	0.147	P	0.323
66	0.033	25	0.1495	Q	0.332
65	0.035	24	0.152	R	0.339
64	0.036	23	0.154	S	0.348
63	0.037	22	0.157	T	0.358
62	0.038	21	0.159	U	0.368
61	0.039	20	0.161	V	0.377
60	0.040	19	0.166	W	0.386
59	0.041	18	0.1695	X	0.397
58	0.042	17	0.173	Y	0.404
57	0.043	16	0.177	Z	0.413

FIGURE 5.31 Drill sizes (American national standard).

Drill	Decimal	Drill	Decimal	Drill	Decimal	Drill	Decimal	Drill	Decimal	Drill	Decimal
.35mm	.0138	1.75mm	.0650	3.70mm	.1457	6.40mm	.2520	9.00mm	.3543	17.00mm	.6693
.38mm	.0150	1.80mm	.0709	3.75mm	.1477	6.50mm	.2559	9.10mm	.3583	17.50mm	.6890
.40mm	.0157	1.85mm	.0728	3.80mm	.1496	6.60mm	.2598	9.20mm	.3622	18.00mm	.7087
.42mm	.0165	1.90mm	.0748	3.90mm	.1535	6.70mm	.2638	9.25mm	.3642	18.50mm	.7283
.45mm	.0177	1.95mm	.0768	4.00mm	.1575	6.75mm	.2658	9.30mm	.3661	19.00mm	.7480
.48mm	.0189	2.00mm	.0787	4.10mm	.1614	6.80mm	.2677	9.40mm	.3701	19.50mm	.7677
.50mm	.0197	2.05mm	.0807	4.20mm	.1654	6.90mm	.2716	9.50mm	.3740	20.00mm	.7874
.55mm	.0217	2.10mm	.0827	4.25mm	.1674	7.00mm	.2756	9.60mm	.3780	20.50mm	.8071
.60mm	.0236	2.15mm	.0846	4.50mm	.1771	7.10mm	.2795	9.70mm	.3819	21.00mm	.8268
.65mm	.0256	2.20mm	.0866	4.60mm	.1811	7.20mm	.2835	9.75mm	.3839	21.50mm	.8465
.70mm	.0276	2.25mm	.0886	4.70mm	.1850	7.25mm	.2855	9.80mm	.3858	22.00mm	.8661
.75mm	.0295	2.30mm	.0905	4.75mm	.1870	7.30mm	.2874	9.90mm	.3898	22.50mm	.8858
.80mm	.0315	2.35mm	.0925	4.80mm	.1890	7.40mm	.2913	10.00mm	.3937	23.00mm	.9055
.85mm	.0335	2.40mm	.0945	4.90mm	.1929	7.50mm	.2953	10.20mm	.4016	23.50mm	.9252
.90mm	.0354	2.45mm	.0965	5.00mm	.1968	7.60mm	.2992	10.50mm	.4134	24.00mm	.9449
.95mm	.0374	2.50mm	.0984	5.10mm	.2008	7.70mm	.3031	10.80mm	.4252	24.50mm	.9646
1.00mm	.0394	2.55mm	.1004	5.20mm	.2047	7.75mm	.3051	11.00mm	.4330	25.00mm	.9843
1.05mm	.0413	2.60mm	.1024	5.25mm	.2067	7.80mm	.3071	11.20mm	.4409		
1.10mm	.0433	2.65mm	.1043	5.30mm	.2087	7.90mm	.3110	11.50mm	.4528		
1.15mm	.0453	2.70mm	.1063	5.40mm	.2126	8.00mm	.3150	11.80mm	.4646	+1.00mm increments up to 48mm.	
1.20mm	.0472	2.75mm	.1083	5.50mm	.2165	8.10mm	.3189	12.00mm	.4724		
1.25mm	.0492	2.80mm	.1102	5.60mm	.2205	8.20mm	.3228	12.20mm	.4803		
1.30mm	.0512	2.90mm	.1142	5.70mm	.2244	8.25mm	.3248	12.50mm	.4921	+5.00mm increments from 50mm up to 105mm.	
1.35mm	.0531	3.00mm	.1181	5.75mm	.2264	8.30mm	.3268	13.00mm	.5118		
1.40mm	.0551	3.10mm	.1220	5.80mm	.2283	8.40mm	.3307	13.50mm	.5315		
1.45mm	.0571	3.20mm	.1260	5.90mm	.2323	8.50mm	.3346	14.00mm	.5512		
1.50mm	.0591	3.25mm	.1280	6.00mm	.2362	8.60mm	.3386	14.50mm	.5709		
1.55mm	.0610	3.30mm	.1299	6.10mm	.2401	8.70mm	.3425	15.00mm	.5906		
1.60mm	.0629	3.40mm	.1339	6.20mm	.2441	8.75mm	.3445	15.50mm	.6102		
1.65mm	.0650	3.50mm	.1378	6.25mm	.2461	8.80mm	.3465	16.00mm	.6299		
1.70mm	.0669	3.60mm	.1417	6.30mm	.2480	8.90mm	.3504	16.50mm	.6496		

FIGURE 5.32 Drill sizes (metric).

Speeds, Feeds, Drill Geometry and Cutting Recommendations for Standard Drill Types *

Material Type	Hardness Bhn	Tool Grade	Drill Type	PA-deg.	LRf deg.	HA-deg.	Point Type	sfpm Speed	ipr Feed x 10⁻³
Low-alloy steels 4130, 4340, 4140	to 300	M-1, M-2	A	118-135	7-10	25-30	Split	50-60	3-7
	300-400	M-1, M-2	A	118-135	7-10	25-30	split	40-50	2-6
	400-500	Cobalt	B	118-135	7-10	25-30	split	25-40	1-4
	over 500	C-2	C	118	7-10	0	notched	75-100	0.5-2
Die steels Hot-work	to 300	M-1, M-2	A	118-135	7-10	25-30	split	45-55	3-7
	300-400	M-1, M-2	A	118-135	7-10	25-30	split	35-50	2-6
	400-500	Cobalt	B	118-135	7-10	25-30	split	25-35	1-4
	over 500	C-2	C	118	7-10	0	Notched	70-90	0.5-2
Stainless steels (Austenitic) 300	135-185	M-1, M-2	A	118-135	7-10	25-30	split	70-90	2-6
Stainless steels (Martensitic) 400	150-250	M-1, M-2	A	118-135	7-10	25-30	split	50-70	3-7
	250-450	M-1, M-2	A	118-135	7-10	25-30	split	30-40	2-6
	over 450	Cobalt	B	118-135	7-10	25-30	split	20-30	1-4
Stainless steels Precipitation hardening 17-7PH, etc.	to 200	M-1, M-2	A	118-135	7-10	25-30	split	50-60	3-7
	200-350	M-1, M-2	A	118-135	7-10	25-30	split	35-45	2-6
	over 350	Cobalt	B	118-135	7-10	25-30	split	20-30	1-4
Nickel-cobalt steels High-strength	to 400	M-1, M-2	A	118-135	7-10	25-30	split	55-65	2-6
	400-500	Cobalt	B	118-135	7-10	25-30	split	30-40	1-4
	over 500	C-2	C	118	7-10	0	notched	70-90	0.5-2
High-temperature Cobalt-base alloys	to 300	Cobalt	B	118-135	7-10	25-30	split	15-25	1-4
High-temperature Iron-base alloys	to 250	Cobalt	B	118-135	7-10	25-30	split	20-30	2-6
	over 250	Cobalt	B	118-135	7-10	25-30	split	15-25	2-6
High-temperature Nickel-base alloys	to 265	Cobalt	B	118-135	7-10	25-30	split	20-30	2-6
	265-330	Cobalt	B	118-135	7-10	25-30	split	20-25	2-5
	over 330	Cobalt	B	118-135	7-10	25-30	split	15-20	1-4

FIGURE 5.33 Drilling recommendation table.

Material								
Magnesium & alloys	All	M-1, M-2	A	118-135	7-10	25-30	split	150-350	2-7
Aluminum & alloys 2024, 6061, 7075, etc.	All	M-1, M-2	A	118-135	7-10	25-30	split	175-400	2-7
Titanium	to 250	M-34, M-42	B	135	7-12	30-38	split	25-30	5-7
Titanium Alpha alloys	250-300	M-34, M-42	B	135	7-12	30-38	split	20-25	5-7
Titanium Alpha-Beta alloys	to 350	M-34, M-42	B	135	7-12	30-38	split	20-25	5-7
	over 350	M-34, M-42	B	135	7-12	30-38	split	15-25	5-7
Titanium Beta alloys	to 350	M-34, M-42	B	135	7-12	30-38	split	15-20	1-4
	over 350	M-34, M-42	B	135	7-12	30-38	split	15-17	0.5-2
Beryllium copper	250	C-2	D	90-118	10-15	25-30	split	30-45	2-8
Tungsten & alloys	to 350	C-2, C-3	D	90-118	7-10	25-30	notched	200-250	1-4
Brass, free-machining	All	M-1, M-2	A	118	7-10	25-30	standard	100-250	4-10
Bronzes, common	All	M-1, M-2	A	118	7-10	25-30	standard	200-250	3-15
Bronze, phosphur	Hard	M-1, M-2	A	118	7-10	25-30	notched	75-150	2-6
Copper	All	M-1, M-2	A	90-118	7-10	25-30	standard	100-250	1-5
Cast iron Soft to medium	soft-med.	M-1, M-2	A	118	7-10	25-30	std or split	75-150	2-8
Cast iron Hard	Hard	C-2	D	118	7-10	25-30	std or split	40-75	1-5
Zinc	All	M-1, M-2	A	118	7-10	25-30	standard	200-250	3-10
Low-carbon steels	to 300	M-1, M-2	A	118	7-10	25-30	standard	80-100	3-10
Thermoplastics	Medium	M-1, M-2	E	60-90	12-16	17	standard	100-150	2-15

FIGURE 5.33 *(Continued)* Drilling recommendation table.

		PA	LRf	HA		ipr		
Thermosetting plastics	Soft	M-1, M-2	E	60-90	12-16	17	standard	150	3-8
	Hard	M-1, M-2	E	60-90	12-16	17	standard	100	2-6

NOTE:
Drill Types: A = AIAA type B or C; B = Heavy duty cobalt; C = Carbide tipped; D = Solid carbide; E = Standard with wide, polished flutes.

*Tabular data in the table is for drills of 0.125 through 0.500" diameter and hole depths of 1 to 3 drill diameters. Adjustments must be made for other conditions, by interpolation or trial drilling. (Smaller drills have a lower ipr feed rate; larger drills have a higher ipr feed rate).

Drill geometry: PA = Point angle, degrees; LRf = Lip relief angle, degrees; HA = helix angle, degrees

Tabular data for ipr - Feed is given in powers of ten notation, i.e. 2 = .002"; 6 = .006"; 0.5 = 0.0005", etc.

FIGURE 5.33 *(Continued)* Drilling recommendation table.

Material	Hardness Bhn	Feed - ipr Spade Drill Diameter, inches					
		1-1.25	1.25-2	2-3	3-4	4-5	5-8
Plain carbon steels	100-225	.012	.015	.018	.022	.025	.030
	225-275	.010	.013	.015	.018	.020	.025
	275-325	.008	.010	.013	.015	.018	.020
Free-machining steels	100-240	.014	.016	.018	.022	.025	.030
	240-325	.010	.014	.016	.020	.022	.025
Free-machining alloy steels	150-250	.014	.016	.018	.022	.025	.030
	250-325	.012	.014	.016	.018	.020	.025
	325-375	.010	.012	.014	.016	.018	.020
Alloy steels	125-180	.012	.015	.018	.022	.025	.030
	180-225	.010	.012	.016	.018	.022	.025
	225-325	.009	.010	.013	.015	.018	.020
	325-400	.006	.006	.010	.012	.014	.016
Grey cast iron	110-160	.020	.022	.026	.028	.030	.034
	160-240	.012	.014	.016	.018	.020	.022
	240-325	.010	.012	.016	.018	.018	.018
Ductile & nodular iron	140-190	.014	.016	.018	.020	.022	.024
	190-250	.012	.014	.016	.018	.018	.020
	250-325	.010	.012	.016	.018	.018	.018
Malleable iron- ferritic	110-160	.014	.016	.018	.020	.022	.024
Malleable iron- pearlitic	160-280	.011	.013	.015	.018	.018	.018
Stainless steels- free-machining Ferritic & austenitic (screw stock) Martensitic (440F, etc.)	——— ———	.016 .012	.016 .014	.018 .014	.020 .016	.022 .016	.024 .018	.026 .020

FIGURE 5.34 Recommended feed rates for spade drilling.

Material	Hardness						
Stainless steels							
Ferritic & austenitic (200 & 300 series)	—	.012	.014	.016	.018	.020	.020
Martensitic (400 series)	—	.010	.012	.012	.014	.016	.018
Copper alloys							
Soft (ETP-110, etc)	—	.016	.018	.020	.026	.028	.030
Hard (bronzes)	—	.010	.012	.014	.016	.018	.018
Aluminum alloys (free-machining)	—	.020	.022	.024	.028	.030	.040
Magnesium alloys (general) **	—	.024	.026	.030	.034	.040	.050
High-temperature alloys (general)	—	.008	.010	.012	.012	.014	.014
Titanium alloys (general)	—	.008	.010	.012	.014	.014	.016
Tool steels							
Water-hardening & shock resisting	150-250	.012	.014	.016	.018	.020	.022
Cold work	200-250	.007	.008	.009	.010	.011	.012
Hot work	150-250	.012	.013	.015	.016	.018	.020
High-speed steels	200-250	.010	.012	.013	.015	.017	.018

NOTE: Hardness ranges are Brinell hardness numbers. Tabular data is in inches of feed per revolution, ipr.
** A fire hazard exists when machining or drilling magnesium & alloys.

FIGURE 5.34 *(Continued)* Recommended feed rates for spade drilling.

Material	Hardness Bhn	"P" Factor
Plain carbon & alloy steels	90-200	0.75
	200-275	0.92
	300-375	1.02
	375-450	1.18
	45-52R_C	1.45
Gray cast irons	-------	0.25
Alloy cast irons & ductile irons	-------	0.50
Stainless steel (austenitic)	-------	0.96
Stainless steels (martensitic)	-------	0.81
Titanium alloys	-------	0.87
Aluminum alloys	-------	0.20
Magnesium alloys	-------	0.15
Copper alloys	Soft - R_B 20-80	0.42
	Hard - R_B 80-100	0.75
Tool steels	-------	1.10
Cobalt based alloys	-------	1.25
High-temperature alloys	-------	1.45
Non-ferrous free-machining alloys	-------	0.45

Note: Where no hardness range is given, the maximum hardness is 300 Bhn. For harder materials, use a higher "P" factor.

FIGURE 5.35 P factor for spade drilling various materials.

$$T_p = 148{,}500 PFD$$

$$M_{hp} = \frac{C_{hp}}{e}$$

$$F = \frac{f_m}{N} \quad \text{and} \quad f_m = FN$$

where C_{hp} = horsepower at the cutter
M_{hp} = required motor horsepower
T_p = thrust for spade drilling, lbf
D = drill diameter, in

5.60   CHAPTER FIVE

F = feed, ipr (see Fig. 5.34 for ipr/diameter/material)
P = power factor constant (see Fig. 5.35)
f_m = feed, ipm
N = spindle speed, rpm
e = drive motor efficiency factor (0.90 for direct belt drive to the spindle; 0.80 for geared head drive to the spindle)

NOTE. The P factors must be increased by 40 to 50 percent for dull tools, although dull cutters should not be utilized if productivity is to remain high.

Problem. Calculate the horsepower at the cutter, required horsepower of the motor, the required thrust force, and the feed in inches per minute, to spade drill carbon steel with a hardness of 275 to 325 Bhn, using a 2.250-in-diameter spade drill rotating at 200 rpm.

Step 1. Find the feed rate for the 2.250-in-diameter drill for the selected carbon steel, from Figure 5.34:

$$F = \text{feed, ipr} = 0.013$$

Step 2. Select the P factor for the material and drill size from Figure 5.35:

$$P = 1.02$$

Step 3. Calculate cutter horsepower:

$$C_{hp} = P\left(\frac{\pi D^2}{4}\right) FN$$

$$C_{hp} = 1.02 \left[\frac{3.1416(2.250)^2}{4}\right] \times 0.013 \times 200$$

$$C_{hp} = 1.02(3.976) \times 0.013 \times 200$$

$$C_{hp} = 10.5 \text{ hp}$$

Step 4. Calculate motor horsepower:

$$M_{hp} = \frac{C_{hp}}{e} = \frac{10.5}{0.95} = 11 \text{ hp at the motor}$$

Step 5. Calculate thrust force:

$$T_p = 148{,}000 PFD$$

$$T_p = 148{,}000 \times 1.02 \times 0.013 \times 2.250$$

$$T_p = 4{,}430 \text{ lbf}$$

Step 6. Calculate feed, ipm:

$$f_m = FN$$

$$f_m = 0.013 \times 200 = 2.60 \text{ ipm}$$

NOTE. If the thrust force cannot be obtained, reduce the feed, ipr, from Fig. 5.34 to a lower value and recalculate the preceding equations. This will lower the horsepower requirement and thrust force, but will also reduce the feed, in/min, taking longer to drill the previously calculated depth per minute, in/min.

5.5 REAMING

A *reamer* is a rotary cutting tool, either cyclindrical or conical in shape, used for enlarging drilled holes to accurate dimensions, normally on the order of ±0.0001 in and closer. Reamers usually have two or more flutes which may be straight or spiral in either left-hand or right hand spiral. Reamers are made for manual or machine operation.

Reamers are made in various forms, including

- Hand reamers
- Machine reamers
- Left-hand flute
- Right-hand flute
- Expansion reamers
- Chucking reamers
- Stub screw-machine reamers
- End-cutting reamers
- Jobbers reamers
- Shell reamers
- Combined drill and reamer

Most reamers are produced from premium-grade HSS. Reamers are also produced in cobalt alloys, and these may be run at speeds 25 percent faster than HSS reamers. Reamer feeds depend on the type of reamer, the material and amount to be removed, and the final finish required. Material-removal rates depend on the size of the reamer and material, but general figures may be used on a trial basis and are summarized here:

Hole diameter	Material to be removed
Up to 0.500 in diameter	0.005 in for finishing
More than 0.500 in diameter	0.015 in for finishing
Up to 0.500 in diameter	0.015 in for semifinished holes
More than 0.500 in diameter	0.030 in for semifinished holes

This is an important consideration when using the expansion reamer owing to the maximum amount of expansion allowed by the adjustment on the expansion reamer.

5.62 CHAPTER FIVE

Machine Speeds and Feeds for HSS Reamers. See Fig. 5.36.

NOTE. Cobalt-alloy and carbide reamers may be run at speeds 25 percent faster than those shown in Fig. 5.36.

Carbide-tipped and solid-carbide chucking reamers are also available and afford greater effective life than HHS and cobalt reamers without losing their nominal size dimensions. Speeds and feeds for carbide reamers are generally similar to those for the cobalt-alloy types.

Forms of Reamers. Other forms of reamers include the following:

Material	Speed (sfpm)	Feed Code (ipr)
Steel - 150 Bhn	80	1
Steel - 200 Bhn	55	2
Steel - 250 Bhn	35	3
Steel - 300 Bhn	30	3
Steel - 350 Bhn	17	4
Steel - 400 Bhn	10	4
Steel, cast	25	3
Steel, forged alloys	30	3
Steel, low carbon	75	2
Steel, high carbon	45	4
Steel, stainless	15	3
Steel, tool	35	4
Titanium	40	1
Zinc alloy	150	1
Aluminum & alloys	150	1
Brass, leaded	175	1
Brass, red & yellow	150	1
Bronzes	160	1
Copper	45	3
Cast iron, chilled	10	4
Cast iron, hard	50	3
Cast iron, pearlitic	60	1
Cast iron, soft	95	1
Malleable iron	65	2
Monels	30	3
Nickels	40	3
Plastic, hard	50	1
Plastic, soft	65	3

Feed Code, ipr (inches per revolution)

Reamer Diameter	Code 1	Code 2	Code 3	Code 4
0.125"	0.006	0.005	0.004	0.003
0.500"	0.012	0.010	0.007	0.005
1.00"	0.020	0.015	0.012	0.008
2.00"	0.032	0.025	0.020	0.012
2.25 - 2.50"	0.043	0.035	0.028	0.018
2.75 - 3.00"	0.055	0.045	0.035	0.024

Note: Reamer feeds may be interpolated for intermediate sizes than those shown in the table. Cobalt reamers may be run at speeds 25% faster than those shown in the table for HSS.

FIGURE 5.36 Machine speeds and feeds for HSS reamers.

Morse taper reamers. These reamers are used to produce and maintain holes for American standard Morse taper shanks. They usually come in a set of two, one for roughing and the other for finishing the tapered hole.

Taper-pin reamers. Taper-pin reamers are produced in HSS with straight, spiral, and helical flutes. They range in size from pin size 7/0 through 14 and include 21 different sizes to accommodate all standard taper pins.

Dowel-pin reamers. Dowel-pin reamers are produced in HSS for standard length and jobbers' lengths in 14 different sizes from 0.125 through 0.500 in. The nominal reamer size is slightly smaller than the pin diameter to afford a force fit.

Helical-flute die-makers' reamers. These reamers are used as milling cutters to join closely drilled holes. They are produced from HSS and are available in 16 sizes ranging from size AAA through O.

Reamer blanks. Reamer blanks are available for use as gauges, guide pins, or punches. They are made of HSS in jobbers' lengths from 0.015- through 0.500-in diameters. Fractional sizes through 1.00-in diameter and wire-gauge sizes are also available.

Shell reamers. These reamers are designed for mounting on arbors and are best suited for sizing and finishing operations. Most shell reamers are produced from HSS. The inside hole in the shell reamer is tapered ⅛ in per foot and fits the taper on the reamer arbor.

Expansion reamers. The hand expansion reamer has an adjusting screw at the cutting end which allows the reamer flutes to expand within certain limits. The recommended expansion limits are listed here for sizes through 1.00-in diameter:

Reamer size: 0.25- to 0.625-in diameter   Expansion limit = 0.010 in
Reamer size: 0.75- to 1.000-in diameter   Expansion limit = 0.013 in

NOTE. Expansion reamer stock sizes up to 3.00-in diameter are available.

5.6 BROACHING

Broaching is a precision machining operation wherein a broach tool is either pulled or pushed through a hole in a workpiece or over the surface of a workpiece to produce a very accurate shape such as round, square, hexagonal, spline, keyway, and so on. Keyways in gear and sprocket hubs are broached to an exact dimension so that the key will fit with very little clearance between the hub of the gear or sprocket and the shaft. The cutting teeth on broaches are increased in size along the axis of the broach so that as the broach is pushed or pulled through the workpiece, a progressive series of cuts is made to the finished size in a single pass.

Broaches are driven or pulled by manual arbor presses and horizontal or vertical broaching machines. A single stroke of the broaching tool completes the machining operation. Broaches are commonly made from premium-quality HSS and are supplied either in single tools or as sets in graduated sizes and different shapes.

Broaches may be used to cut internal or external shapes on workpieces. Blind holes also can be broached with specially designed broaching tools. The broaching tool teeth along the length of the broach are normally divided into three separate sections. The teeth of a broach include roughing teeth, semifinishing teeth, and finishing teeth. All finishing teeth of a broach are the same size, while the semifinishing and roughing teeth are progressive in size up to the finishing teeth.

A broaching tool must have sufficient strength and stock-removal and chip-carrying capacity for its intended operation. An interval-pull broach must have sufficient tensile strength to withstand the maximum pulling forces that occur during the pulling operation. An internal-push broach must have sufficient compressive strength as well as the ability to withstand buckling or breaking under the pushing forces that occur during the pushing operation.

Broaches are produced in sizes ranging from 0.050 in to as large as 20 in or more. The term *button broach* is used for broaching tools which produce the spiral lands that form the rifling in gun barrels from small to large caliber. Broaches may be rotated to produce a predetermined spiral angle during the pull or push operations.

Calculation of Pull Forces During Broaching. The allowable pulling force P is determined by first calculating the cross-sectional area at the minimum root of the broach. The allowable pull in pounds force is determined from

$$P = \frac{A_r F_y}{f_s}$$

where A_r = minimum tool cross section, in^2
F_y = tensile yield strength or yield point of tool steel, psi
f_s = factor of safety (generally 3 for pull broaching)

The minimum root cross section for a round broach is

$$A_r = \frac{\pi D_r^2}{4} \quad \text{or} \quad 0.7854 D_r^2$$

where D_r = minimum root diameter, in
The minimum pull-end cross section A_p is

$$A_p = \frac{\pi}{4} D_p^2 - W D_p \quad \text{or} \quad 0.7854 D_p^2 - W D_p$$

where D_p = pull-end diameter, in
W = pull-slot width, in

Calculation of Push Forces During Broaching. Knowing the length L and the compressive yield point of the tool steel used in the broach, the following relations may be used in designing or determining the maximum push forces allowed in push broaching.

If the length of the broach is L and the minimum tool diameter is D_r, the ratio L/D_r should be less than 25 so that the tool will not bend under maximum load. Most push broaches are short enough that the maximum compressive strength of the broach material will allow much greater forces than the forces applied during the broaching operation.

FORMULAS AND CALCULATIONS FOR MACHINING OPERATIONS   5.65

If the L/D_r ratio is greater than 25, compressive broaching forces may bend or break the broach tool if they exceed the maximum allowable force for the tool. The maximum allowable compressive force (pounds force) for a long push broach is determined from the following equation:

$$P = \frac{5.6 \times 10^7 D_r^4}{(f_s) L^2}$$

where L is measured from the push end to the first tooth in inches.

Minimum Forces Required for Broaching Different Materials. For flat-surface broaches,

$$F = WnR\psi$$

For round-hole internal broaches,

$$F = \frac{\pi D n R}{2} \psi$$

For spline-hole broaches,

$$F = \frac{nSWR}{2} \psi$$

where F = minimum pulling or pushing force required, lbf
        W = width of cut per tooth or spline, in
        D = hole diameter before broaching, in
        R = rise per tooth, in
        n = maximum number of broach teeth engaged in the workpiece
        S = number of splines (for splined holes only)
        ψ = broaching constant (see Fig. 5.37 for values)

Material	Value of ψ
Aluminum	200,000 - 300,000
Babbitt	25,000 - 35,000
Brass	200,000 - 300,000
Bronze	300,000 - 350,000
Cast irons	200,000 - 350,000
High-temperature alloys	350,000 - 600,000
Mild steels	350,000 - 450,000
Steel castings	350,000 - 400,000
Titanium	325,000 - 375,000
Zinc alloys	200,000 - 250,000

Note: The tabular values given in the table have a limited value due to the many variables involved in broaching, such as chipbreakers, lubricating and cutting fluid effects and other factors which tend to increase or reduce the required cutting force as calculated using the preceding equations.

FIGURE 5.37   Broaching constants ψ for various materials.

Problem. You need to push-broach a 0.625-in-square hole through a 0.3125-in-thick bar made of C-1018 mild steel. Your square broach has a rise per tooth R of 0.0035 in and a tooth pitch of 0.250 in.

Solution. Use the following equation (shown previously for flat-surface broaches). Before broaching, drill a hole through the bar using a $\frac{41}{64}$-in-diameter drill, or a drill which is 0.015 to 0.20 in larger than a side of the square hole.

$$F = WnR\psi$$

where $W = 0.625$ in (side of square)
$n = 4$ sides $\times$ 2 rows in contact = 8 (maximum teeth in engagement)
$R = 0.0035$ in, given or measured on the broach
$\psi = 400{,}000$ (mean value given in Fig. 5.37 for mild steel)
$F = $ maximum force, lb, required on the broach, lbf

So,

$$F = 0.625 \times 8 \times 0.0035 \times 400{,}000$$

$F = 7000$ lbf   (maximum push force on the broach)

Now, measure the root diameter D_r and length L of the broach, and use a factor of safety f_s of 2. Then check to see if your broach can withstand the 7000-lb push force P required to broach the hole:

$$P = \frac{5.6 \times 10^7 (D_r)^4}{(f_s)L^2}$$

If the root diameter D_r of the square broach = 0.500 in, and the effective broach length $L = 14$ in, then:

$$P = \frac{5.6 \times 10^7 (0.500)^4}{2(14)^2}$$

$$P = \frac{3{,}500{,}000}{392} = 8929 \text{ lbf}$$ (allowed push on the broach)

The calculations indicate that the square broach described will withstand the 7000-lb push, even though its L/D_r ratio is 14/0.500 = 28, which is greater than the ratio of 25. We used the preceding equation because the broach L/D_r ratio was greater than 25, and we considered it a *long broach,* requiring the use of this equation. If the L/D_r ratio of the broach is less than 25, the use of this equation is normally not required.

5.7 VERTICAL BORING AND JIG BORING

The increased demand for accuracy in producing large parts initiated the refined development of modern vertical and jig boring machines. Although the modern CNC machining centers can handle small to medium-sized jig boring operations, very large

FORMULAS AND CALCULATIONS FOR MACHINING OPERATIONS 5.67

and heavy work of high precision is done on modern CNC jig boring machines or vertical boring machines. Also, any size work which requires extreme accuracy is usually jig bored. The modern jig boring machines are equipped with high-precision spindles and *x/y* coordinate table movements of high precision and may be CNC machines with digital read-out panels. For a modern CNC/DNC jig boring operation, the circle diameter and number of equally spaced holes or other geometric pattern is entered into the DNC program and the computer calculates all the coordinates and orientation of the holes from a reference point. This information is either sent to the CNC jig boring machine's controller or the machine operator can load this information into the controller, which controls the machine movements to complete the machining operation.

Extensive tables of jig boring coordinates are not necessary with the modern CNC jig boring or vertical boring machines. Figures 5.38 and 5.39 are for manually controlled machines, where the machine operator makes the movements and coordinate settings manually.

Vertical boring machines with tables up to 192 in in diameter are produced for machining very large and heavy workpieces.

For manually controlled machines with vernier or digital readouts, a table of jig boring dimensional coordinates is shown in Fig. 5.38 for dividing a 1-in circle into a number of equal divisions. Since the dimensions or coordinates given in the table are for *xy* table movements, the machine operator may use these directly to make the appropriate machine settings after converting the coordinates for the required circle diameter to be divided.

Figure 5.39 is a coordinate diagram of a jig bore layout for 11 equally spaced holes on a 1-in-diameter circle. The coordinates are taken from the table in Fig. 5.38. If a different-diameter circle is to be divided, simply multiply the coordinate values in the table by the diameter of the required circle; i.e., for an 11-hole circle of 5-in diameter, multiply the coordinates for the 11-hole circle by 5. Thus the first hole *x* dimension would be $5 \times 0.50000 = 2.50000$ in, and so on. Figure 5.40 shows a typical boring head for removable inserts.

5.8 BOLT CIRCLES (BCS) AND HOLE COORDINATE CALCULATIONS

This covers calculating the hole coordinates when the bolt circle diameter and angle of the hole is given. Refer to Fig. 5.41, where we wish to find the coordinates of the hole in quadrant II, when the bolt circle diameter is 4.75 in, and the angle given is 37.5184°.

The radius R is therefore 2.375 in, and we can proceed to find the *x* or horizontal ordinate from:

$$\cos 37.5184 = \frac{H}{R}$$

$$H = 2.375 \times \cos 37.5184$$

$$H = 2.375 \times 0.7932$$

$$H = 1.8839 \text{ in}$$

Hole No.	Horizontal X	Vertical Y	Hole No.	Horizontal X	Vertical Y
Three holes:			Thirteen holes:		
1	0.50000		1	0.50000	
2	0.75000	0.43301	2	0.05727	0.23236
3	----------	0.86602	3, 13	0.15870	0.17913
Five holes:			4, 12	0.22376	0.08486
1	0.50000		5, 11	0.23757	0.02885
2	0.34549	0.47553	6, 10	0.19695	0.13594
3, 5	0.55902	0.18164	7, 9	0.11121	0.21190
4	----------	0.58778	8	----------	0.23932
Six holes:			Fourteen holes:		
1, 3, 6	0.50000		1	0.50000	
2, 4, 5	0.25000	0.43301	2, 8, 9	0.04951	0.21694
Seven holes:			3, 7, 10, 14	0.13875	0.17397
1	0.50000		4, 6, 11, 13	0.20048	0.09655
2	0.18826	0.39091	5, 12	0.22252	
3, 7	0.42300	0.09655	Fifteen holes:		
4, 6	0.33923	0.27052	1	0.50000	
5	----------	0.43388	2	0.04323	0.20337
Eight holes:			3, 15	0.12221	0.16820
1	0.50000		4, 14	0.18005	0.10396
2, 5, 6	0.14645	0.35355	5, 13	0.20677	0.02173
3, 4, 7, 8	0.35355	0.14645	6, 12	0.19774	0.06425
Nine holes:			7, 11	0.15451	0.13912
1	0.50000		8, 10	0.08456	0.18994
2	0.11698	0.32139	9	----------	0.20790
3, 9	0.29620	0.17101	Sixteen holes:		
4, 8	0.33682	0.05939	1	0.50000	
5, 7	0.21084	0.26200	2, 9, 10	0.03806	0.19134
6	----------	0.34202	3, 8, 11, 16	0.10839	0.16221
Ten holes:			4, 7, 12, 15	0.16221	0.10839
1	0.50000		5, 6, 13, 14	0.19134	0.03806
2, 6, 7	0.09549	0.29389	Seventeen holes:		
3, 5, 8, 10	0.25000	0.18164	1	0.50000	
4, 9	0.30902		2	0.03377	0.18062
Eleven holes:			3, 17	0.09672	0.15623
1	0.50000		4, 16	0.14664	0.11073
2	0.07937	0.27032	5, 15	0.17674	0.05028
3, 11	0.21292	0.18450	6, 14	0.18296	0.01695
4, 10	0.27887	0.04009	7, 13	0.16449	0.08190
5, 9	0.25626	0.11704	8, 12	0.12379	0.13580
6, 8	0.15233	0.23701	9, 11	0.06637	0.17134
7	----------	0.28172	10	----------	0.18374
Twelve holes:			Eighteen holes:		
1	0.50000		1	0.50000	
2, 7, 8	0.06699	0.25000	2, 10, 11	0.03016	0.17101
3, 6, 9, 12	0.18301	0.18301	3, 9, 12, 18	0.08682	0.15038
4, 5, 10, 11	0.25000	0.06699	4, 8, 13, 17	0.13302	0.11162
			5, 7, 14, 16	0.16318	0.05939
			6, 15	0.17364	

FIGURE 5.38 Jig-boring coordinates for dividing the circle.

FORMULAS AND CALCULATIONS FOR MACHINING OPERATIONS 5.69

Hole No.	Horizontal X	Vertical Y	Hole No.	Horizontal X	Vertical Y
Nineteen holes:			11, 15	0.07076	0.11634
1	0.50000		12, 14	0.03673	0.13112
2	0.02709	0.16235	13	----------	0.13616
3, 10	0.07834	0.14475	Twenty-four holes:		
4, 18	0.12110	0.11148	1	0.50000	
5, 17	0.15073	0.06612	2, 13, 14	0.01704	0.12941
6, 16	0.16403	0.01358	3, 12, 15, 24	0.04995	0.12059
7, 15	0.15956	0.04039	4, 11, 16, 23	0.07946	0.10355
8, 14	0.13779	0.09003	5, 10, 17, 22	0.10355	0.07946
9, 13	0.10110	0.12989	6, 9, 18, 21	0.12059	0.04995
10, 12	0.05344	0.15567	7, 8, 19, 20	0.12941	0.01704
11	----------	0.16460	Twenty-five holes:		
Twenty holes:			1	0.50000	
1	0.50000		2	0.01508	0.12434
2, 11, 12	0.02447	0.15451	3, 25	0.04677	0.11653
3, 10, 13, 20	0.07102	0.13938	4, 24	0.07367	0.10140
4, 9, 14, 19	0.11062	0.11062	5, 23	0.09657	0.07989
5, 8, 15, 18	0.13938	0.07102	6, 22	0.11340	0.05337
6, 7, 16, 17	0.15451	0.02447	7, 21	0.12312	0.02348
Twenty-one holes:			8, 20	0.12508	0.00787
1	0.50000		9, 19	0.11920	0.03873
2	0.02221	0.14738	10, 15	0.10582	0.06716
3, 21	0.06467	0.13428	11, 17	0.08580	0.09136
4, 20	0.10138	0.10925	12, 16	0.06038	0.10983
5, 19	0.12908	0.07452	13, 15	0.03116	0.12140
6, 18	0.14530	0.03317	14	----------	0.12532
7, 17	0.14862	0.01114	Twenty-six holes:		
8, 16	0.13874	0.05445	1	0.50000	
9, 15	0.11652	0.09293	2, 14, 15	0.01454	0.11966
10, 14	0.08397	0.12314	3, 13, 16, 26	0.04273	0.11270
11, 13	0.04393	0.14242	4, 12, 17, 25	0.06848	0.09920
12	----------	0.14904	5, 11, 18, 24	0.09022	0.07993
Twenty-two holes:			6, 10, 19, 23	0.10673	0.05601
1	0.50000		7, 9, 20, 22	0.11703	0.02885
2, 12, 13	0.02025	0.14086	8, 21	0.12054	
3, 11, 14, 22	0.05912	0.12946	Twenty-seven holes		
4, 10, 15, 21	0.09321	0.10755	1	0.50000	
5, 9, 16, 20	0.11971	0.07695	2	0.01348	0.11530
6, 8, 17, 19	0.13655	0.04009	3, 27	0.03971	0.10910
7, 18	0.14232		4, 26	0.06379	0.09699
Twenty-three hole			5, 25	0.08444	0.07967
1	0.50000		6, 24	0.10054	0.05805
2	0.01854	0.13490	7, 23	0.11121	0.03329
3, 23	0.05425	0.12480	8, 22	0.11580	0.00675
4, 22	0.08593	0.10562	9, 21	0.11433	0.02016
5, 21	0.11125	0.07853	10, 20	0.10660	0.04598
6, 20	0.12830	0.04560	11, 19	0.09312	0.06933
7, 19	0.13585	0.00930	12, 18	0.07462	0.08893
8, 18	0.13331	0.02771	13, 17	0.05210	0.10374
9, 17	0.12091	0.06264	14, 16	0.02678	0.11297
10, 16	0.09951	0.09295	15	----------	0.11608

FIGURE 5.38 *(Continued)* Jig-boring coordinates for dividing the circle.

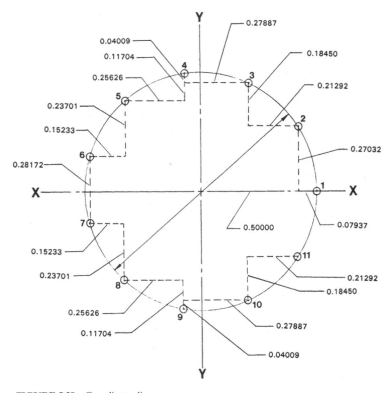

FIGURE 5.39 Coordinate diagram.

FIGURE 5.40 A modern removable insert boring head.

FORMULAS AND CALCULATIONS FOR MACHINING OPERATIONS 5.71

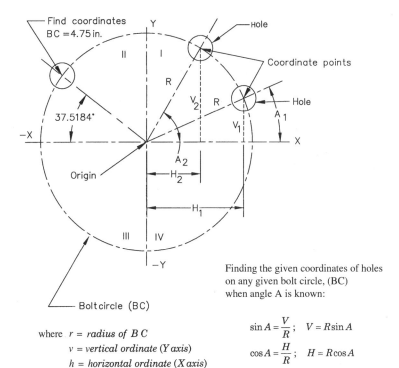

where r = radius of BC
v = vertical ordinate (Y axis)
h = horizontal ordinate (X axis)
a = angle of hole from X axis

Finding the given coordinates of holes on any given bolt circle, (BC) when angle A is known:

$$\sin A = \frac{V}{R}; \quad V = R \sin A$$

$$\cos A = \frac{H}{R}; \quad H = R \cos A$$

FIGURE 5.41 Bolt circle and coordinate calculations.

The y or vertical ordinate is then found from:

$$\sin 37.5184 = \frac{V}{R}$$

$$V = 2.375 \times \sin 37.5184°$$

$$V = 2.375 \times 0.6090$$

$$V = 1.4464 \text{ in}$$

Therefore, the x dimension = 1.8839 in, and the y dimension = 1.4464 in. We can check these answers by using the pythagorean theorem:

$$R^2 = x^2 + y^2$$

$$2.375^2 = 1.8839^2 + 1.4464^2$$

$$5.6406 = 3.5491 + 2.0921$$

$$5.641 = 5.641 \quad \text{(showing an equality accurate to 3 decimal places)}$$

5.72 CHAPTER FIVE

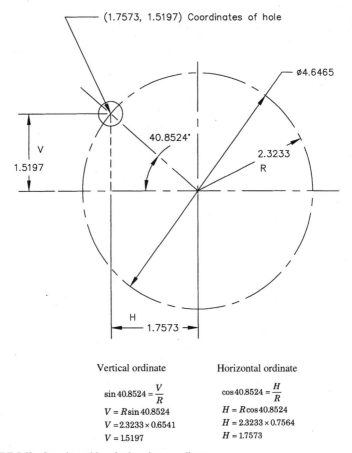

FIGURE 5.42 Sample problem for locating coordinates.

Figure 5.42 shows another sample calculation for obtaining the coordinates of a hole at 40.8524° on a bolt circle with a diameter of 4.6465 in.

CHAPTER 6
FORMULAS FOR SHEET METAL LAYOUT AND FABRICATION

The branch of metalworking known as *sheet metal* comprises a large and important element. Sheet metal parts are used in countless commercial and military products. Sheet metal parts are found on almost every product produced by the metalworking industries throughout the world.

Sheet metal gauges run from under 0.001 in to 0.500 in. Hot-rolled steel products can run from ½ in thick to no. 18 gauge (0.0478 in) and still be considered sheet. Cold-rolled steel sheets are generally available from stock in sizes from 10 gauge (0.1345 in) down to 28 gauge (0.0148 in). Other sheet thicknesses are available as special-order "mill-run" products when the order is large enough. Large manufacturers who use vast tonnages of steel products, such as the automobile makers, switch-gear producers, and other sheet metal fabricators, may order their steel to their own specifications (composition, gauges, and physical properties).

The steel sheets are supplied in flat form or rolled into coils. Flat-form sheets are made to specific standard sizes unless ordered to special nonstandard dimensions.

The following sections show the methods used to calculate flat patterns for brake-bent or die-formed sheet metal parts. The later sections describe the geometry and instructions for laying out sheet metal developments and transitions. Also included are calculations for punching requirements of sheet metal parts and tooling requirements for punching and bending sheet metals.

Tables of sheet metal gauges and recommended bend radii and shear strengths for different metals and alloys are shown also.

The designer and tool engineer should be familiar with all machinery used to manufacture parts in a factory. These specialists must know the limitations of the machinery that will produce the parts as designed and tooled. Coordination of design with the tooling and manufacturing departments within a company is essential to the quality and economics of the products that are manufactured. Modern machinery has been designed and is constantly being improved to allow the manufacture of a quality product at an affordable price to the consumer. Medium- to large-sized companies can no longer afford to manufacture products whose quality standards do not meet the demands and requirements of the end user.

Modern Sheet Metal Manufacturing Machinery. The processing of sheet metal begins with the hydraulic shear, where the material is squared and cut to size for the next operation. These types of machines are the workhorses of the typical sheet metal department, since all operations on sheet metal parts start at the shear.

Figure 6.1 shows a Wiedemann Optishear, which shears and squares the sheet metal to a high degree of accuracy. Blanks which are used in blanking, punching, and forming dies are produced on this machine, as are other flat and accurate pieces which proceed to the next stage of manufacture.

FIGURE 6.1 Sheet metal shear.

The flat, sheared sheet metal parts may then be routed to the punch presses, where holes of various sizes and patterns are produced. Figure 6.2 shows a medium-sized computer numerically controlled (CNC) multistation turret punch press, which is both highly accurate and very high speed.

Many branches of industry use large quantities of sheet steels in their products. The electrical power distribution industries use very large quantities of sheet steels in 7-, 11-, 13-, and 16-gauge thicknesses. A lineup of electrical power distribution switchgear is shown in Fig. 6.3; the majority of the sheet metal is 11 gauge (0.1196 in thick).

Gauging Systems. To specify the thickness of different metal products, such as steel sheet, wire, strip, tubing, music wire, and others, a host of gauging systems were developed over the course of many years. Shown in Fig. 6.4 are the common gauging systems used for commercial steel sheet, strip, and tubing and brass and steel wire.

FORMULAS FOR SHEET METAL LAYOUT AND FABRICATION

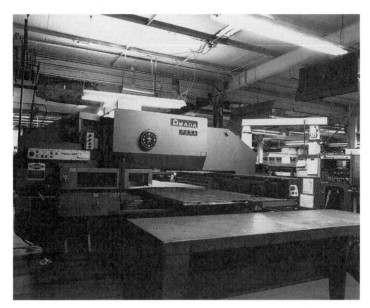

FIGURE 6.2 CNC multistation turret punch press.

FIGURE 6.3 Industrial equipment made from sheet metal.

Gauge No.	Brass (Brown & Sharpe)	Steel Sheets *	Strip & Tubing	Steel Wire Ga. ♦
6-0	0.5800	--------	--------	0.4615
5-0	0.5165	--------	0.500	0.4305
4-0	0.4600	--------	0.454	0.3938
3-0	0.4096	--------	0.425	0.3625
2-0	0.3648	--------	0.380	0.3310
0	0.3249	--------	0.340	0.3065
1	0.2893	--------	0.300	0.2830
2	0.2576	--------	0.284	0.2625
3	0.2294	0.2391	0.259	0.2437
4	0.2043	0.2242	0.238	0.2253
5	0.1819	0.2092	0.220	0.2070
6	0.1620	0.1943	0.203	0.1920
7	0.1443	0.1793	0.180	0.1770
8	0.1285	0.1644	0.165	0.1620
9	0.1144	0.1495	0.148	0.1483
10	0.1019	0.1345	0.134	0.1350
11	0.0907	0.1196 ✓	0.120	0.1205
12	0.0808	0.1046	0.109	0.1055
13	0.0720	0.0897	0.095	0.0915
14	0.0641	0.0747	0.083	0.0800
15	0.0571	0.0673	0.072	0.0720
16	0.0508	0.0598	0.065	0.0625
17	0.0453	0.0538	0.058	0.0540
18	0.0403	0.0478	0.049	0.0475
19	0.0359	0.0418	0.042	0.0410
20	0.0320	0.0359	0.035	0.0348
21	0.0285	0.0329	0.032	0.0317
22	0.0253	0.0299	0.028	0.0286
23	0.0226	0.0269	0.025	0.0258
24	0.0201	0.0239	0.022	0.0230
25	0.0179	0.0209	0.020	0.0204
26	0.0159	0.0179	0.018	0.0181
27	0.0142	0.0164	0.016	0.0173
28	0.0126	0.0149	0.014	0.0162
29	0.0113	0.0135	0.013	0.0150
30	0.0100	0.0120	0.012	0.0140
31	0.0089	0.0105	0.010	0.0132
32	0.0080	0.0097	0.009	0.0128
33	0.0071	0.0090	0.008	0.0118
34	0.0063	0.0082	0.007	0.0104
35	0.0056	0.0075	0.005	0.0095
36	0.0050	0.0067	0.004	0.0090
37	0.0045	0.0064	--------	0.0085
38	0.0040	0.0060	--------	0.0080

*= Common Commercial Standard; ♦ = Reference only

FIGURE 6.4 Modern gauging system chart.

The steel sheets column in Fig. 6.4 lists the gauges and equivalent thicknesses used by American steel sheet manufacturers and steelmakers. This gauging system can be recognized immediately by its 11-gauge equivalent of 0.1196 in, which is standard today for this very common and high-usage gauge of sheet steel.

Figure 6.5 shows a table of gauging systems that were used widely in the past, although some are still in use today, including the American or Brown and Sharpe system. The Brown and Sharpe system is also shown in Fig. 6.4, but there it is indicated in only four-place decimal equivalents.

Figure 6.6 shows weights versus thicknesses of steel sheets.

Number of wire gauge	American or Brown & Sharpe	Birmingham or Stubs' Iron wire	Washburn & Moen, Worcester, Mass.	W & M steel music wire	American S & W Co. music wire gauge	Stubs' steel wire	U.S. standard gauge for sheet and plate iron and steel	Number of wire gauge
00000000				0.0083				00000000
0000000				0.0087				0000000
000000				0.0095	0.004		0.46875	000000
00000				0.010	0.005		0.4375	00000
0000	0.460	0.454	0.3938	0.011	0.006		0.40625	0000
000	0.40964	0.425	0.3625	0.012	0.007		0.375	000
00	0.3648	0.380	0.3310	0.0133	0.008		0.34375	00
0	0.32486	0.340	0.3065	0.0144	0.009		0.3125	0
1	0.2893	0.300	0.2830	0.0156	0.010	0.227	0.28125	1
2	0.25763	0.284	0.2625	0.0166	0.011	0.219	0.265625	2
3	0.22942	0.259	0.2437	0.0178	0.012	0.212	0.25	3
4	0.20431	0.238	0.2253	0.0188	0.013	0.207	0.234375	4
5	0.18194	0.220	0.2070	0.0202	0.014	0.204	0.21875	5
6	0.16202	0.203	0.1920	0.0215	0.016	0.201	0.203125	6
7	0.14428	0.180	0.1770	0.023	0.018	0.199	0.1875	7
8	0.12849	0.165	0.1620	0.0243	0.020	0.197	0.171875	8
9	0.11443	0.148	0.1483	0.0256	0.022	0.194	0.15625	9
10	0.10189	0.134	0.1350	0.027	0.024	0.191	0.140625	10
11	0.090742	0.120	0.1205	0.0284	0.026	0.188	0.125	11
12	0.080808	0.109	0.1055	0.0296	0.029	0.185	0.109375	12
13	0.071961	0.095	0.0915	0.0314	0.031	0.182	0.09375	13
14	0.064084	0.083	0.0800	0.0326	0.033	0.180	0.078125	14

FIGURE 6.5 Early gauging systems.

15	0.057068	0.072	0.0720	0.0345	0.035	0.178	0.0703125	15
16	0.05082	0.065	0.0625	0.036	0.037	0.175	0.0625	16
17	0.045257	0.058	0.0540	0.0377	0.039	0.172	0.05625	17
18	0.040303	0.049	0.0475	0.0395	0.041	0.168	0.050	18
19	0.03589	0.042	0.0410	0.0414	0.043	0.164	0.04375	19
20	0.031961	0.035	0.0348	0.0434	0.045	0.161	0.0375	20
21	0.028462	0.032	0.03175| 0.046	0.047	0.157	0.034375	21
22	0.026347	0.028	0.0286	0.0483	0.049	0.155	0.03125	22
23	0.022571	0.025	0.0258	0.051	0.051	0.153	0.028125	23
24	0.0201	0.022	0.0230	0.055	0.055	0.151	0.025	24
25	0.0179	0.020	0.0204	0.0586	0.059	0.148	0.021875	25
26	0.01594	0.018	0.0181	0.0626	0.063	0.146	0.01875	26
27	0.014195	0.016	0.0173	0.0658	0.067	0.143	0.0171875	27
28	0.012641	0.014	0.0162	0.072	0.071	0.139	0.015625	28
29	0.011257	0.013	0.0150	0.076	0.075	0.134	0.0140625	29
30	0.010025	0.012	0.0140	0.080	0.080	0.127	0.0125	30
31	0.008928	0.010	0.0132		0.085	0.120	0.0109375	31
32	0.00795	0.009	0.0128		0.090	0.115	0.01015625| 32
33	0.00708	0.008	0.0118		0.095	0.112	0.009375	33
34	0.006304	0.007	0.0104			0.110	0.00859375| 34
35	0.005614	0.005	0.0095			0.108	0.0078125	35
36	0.005	0.004	0.0090			0.106	0.00703125| 36
37	0.004453					0.103	0.006640625| 37
38	0.003965					0.101	0.00625	38
39	0.003531					0.099		39
40	0.003144					0.097		40

FIGURE 6.5 (*Continued*) Early gauging systems.

FORMULAS FOR SHEET METAL LAYOUT AND FABRICATION   6.7

Standard gauge number	Weight, oz/ft^2	Weight, lb/ft^2	Thickness, in
3	160	10.0000	0.2391
4	150	9.3750	0.2242
5	140	8.7500	0.2092
6	130	8.1250	0.1943
7	120	7.5000	0.1793
8	110	6.8750	0.1644
9	100	6.2500	0.1495
10	90	5.6250	0.1345
11	80	5.0000	0.1196
12	70	4.3750	0.1046
13	60	3.7500	0.0897
14	50	3.1250	0.0747
15	45	2.8125	0.0673
16	40	2.5000	0.0598
17	36	2.2500	0.0538
18	32	2.0000	0.0478
19	28	1.7500	0.0418
20	24	1.5000	0.0359
21	22	1.3750	0.0329
22	20	1.2500	0.0299
23	18	1.1250	0.0269
24	16	1.0000	0.0239
25	14	0.87500	0.0209
26	12	0.75000	0.0179
27	11	0.68750	0.0164
28	10	0.62500	0.0149
29	9	0.56250	0.0135
30	8	0.50000	0.0120
31	7	0.43750	0.0105
32	6.5	0.40625	0.0097
33	6	0.37500	0.0090
34	5.5	0.34375	0.0082
35	5	0.31250	0.0075
36	4.5	0.28125	0.0067
37	4.25	0.26562	0.0064
38	4	0.25000	0.0060

FIGURE 6.6   Standard gauges and weights of steel sheets.

Aluminum Sheet Metal Standard Thicknesses. Aluminum is used widely in the aerospace industry, and over the years, the gauge thicknesses of aluminum sheets have developed on their own. Aluminum sheet is now generally available in the thicknesses shown in Fig. 6.7. The fact that the final weight of an aerospace vehicle is very critical to its performance has played an important role in the development of the standard aluminum sheet gauge sizes.

Standard Thickness, in.	Weight, lbs/sq. ft.
0.010	0.141
0.016	0.226
0.020	0.282
0.025	0.353
0.032	0.452
0.040	0.564
0.050	0.706
0.063	0.889
0.071	1.002
0.080	1.129
0.090	1.270
0.100	1.411
0.125	1.764
0.160	2.258
0.190	2.681
0.250	3.528

Weight based on an average aluminum weight of 0.098 lb/in^3

FIGURE 6.7 Standard aluminum sheet metal thicknesses and weights.

6.1 SHEET METAL FLAT-PATTERN DEVELOPMENT AND BENDING

The correct determination of the flat-pattern dimensions of a sheet metal part which is formed or bent is of prime importance to sheet metal workers, designers, and design drafters. There are three methods for performing the calculations to determine flat patterns which are considered normal practice. The method chosen also can determine the accuracy of the results. The three common methods employed for doing the work include

1. By bend deduction (BD) or setback
2. By bend allowance (BA)
3. By inside dimensions (IML), for sharply bent parts only

Other methods are also used for calculating the flat-pattern length of sheet metal parts. Some take into consideration the ductility of the material, and others are based on extensive experimental data for determining the bend allowances. The methods included in this section are accurate when the bend radius has been selected properly for each particular gauge and condition of the material. When the proper bend radius is selected, there is no stretching of the *neutral axis* within the part (the neutral axis is generally accepted as being located 0.445 × material thickness inside the inside mold line [IML]) (see Figs. 6.8a and b for calculations, and also see Fig. 6.9).

Methods of Determining Flat Patterns. Refer to Fig. 6.9.
   Method 1. By bend deduction or setback:

$$L = a + b - \text{setback}$$

FORMULAS FOR SHEET METAL LAYOUT AND FABRICATION   6.9

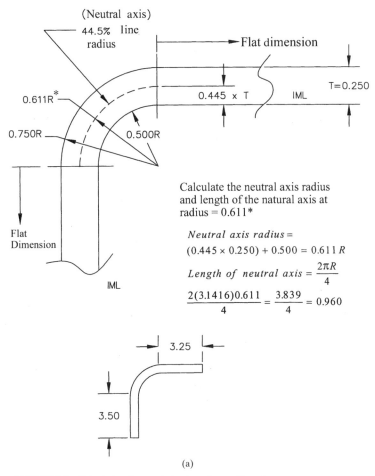

FIGURE 6.8 Calculating the neutral axis radius and length.

Method 2. By bend allowance:

$$L = a' + b' + c$$

where c = bend allowance or length along neutral axis (see Fig. 6.9).

Method 3.   By inside dimensions or inside mold line (IML):

$$L = (a - T) + (b - T)$$

The calculation of bend allowance and bend deduction (setback) is keyed to Fig. 6.10 and is as follows:

6.10 CHAPTER SIX

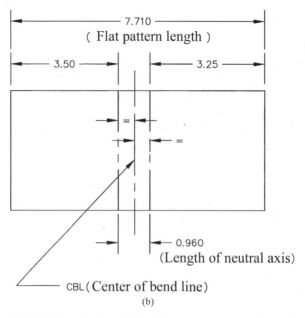

FIGURE 6.8 *(Continued)* Calculating the neutral axis radius and length.

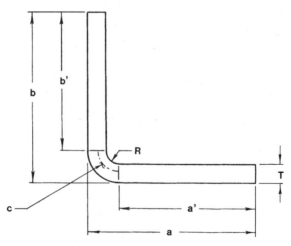

FIGURE 6.9 Bend allowance by neutral axis *c*.

FORMULAS FOR SHEET METAL LAYOUT AND FABRICATION 6.11

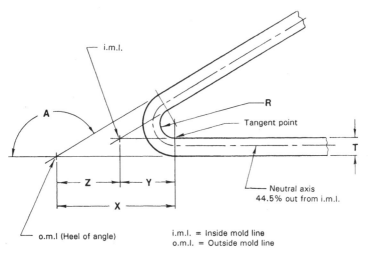

FIGURE 6.10 Bend allowance and deduction.

Bend allowance (BA) = $A(0.01745R + 0.00778T)$

Bend deduction (BD) = $\left(2 \tan \dfrac{1}{2} A\right)(R + T) - (\text{BA})$

$$X = \left(\tan \dfrac{1}{2} A\right)(R + T)$$

$$Z = T\left(\tan \dfrac{1}{2} A\right)$$

$Y = X - Z$ or $R\left(\tan \dfrac{1}{2} A\right)$

On "open" angles that are bent less than 90° (see Fig. 6.11),

$$X = \left(\tan \dfrac{1}{2} A\right)(R + T)$$

Setback or J Chart for Determining Bend Deductions. Figure 6.12 shows a form of bend deduction (BD) or setback chart known as a *J chart*. You may use this chart to determine bend deduction or setback when the angle of bend, material thickness, and inside bend radius are known. The chart in the figure shows a sample line running from the top to the bottom and drawn through the ³⁄₁₆-in radius and the material thickness of 0.075 in. For a 90° bend, read across from the right to where the line intersects the closest curved line in the body of the chart. In this case, it can be seen that the line

6.12 CHAPTER SIX

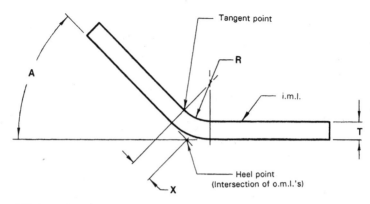

FIGURE 6.11 Open angles less than 90 degrees.

intersects the curve whose value is 0.18. This value is then the required setback or bend deduction for a bend of 90° in a part whose thickness is 0.075 in with an inside bend radius of ³⁄₁₆ in. If we check this setback or bend deduction value using the appropriate equations shown previously, we can check the value given by the J chart.

Checking. Bend deduction (BD) or setback is given as

$$\text{Bend deduction or setback} = \left(2\tan\frac{1}{2}A\right)(R+T) - (\text{BA})$$

We must first find the bend allowance from

$$\text{Bend allowance} = A(0.01745R + 0.00778T)$$

$$= 90(0.01745 \times 0.1875 + 0.00778 \times 0.075)$$

$$= 90(0.003855)$$

$$= 0.34695$$

Now, substituting the bend allowance of 0.34695 into the bend deduction equation yields

$$\text{Bend deduction or setback} = \left[2\tan\frac{1}{2}(90)\right](0.1875 + 0.075) - 0.34695$$

$$= (2 \times 1)(0.2625) - 0.34695$$

$$= 0.525 - 0.34695$$

$$= 0.178 \text{ or } 0.18, \text{ as shown in the chart (Fig. 6.12)}$$

The J chart in Fig. 6.12 is thus an important tool for determining the bend deduction or setback of sheet metal flat patterns without recourse to tedious calculations. The accuracy of this chart has been shown to be of a high order. This chart as well as

FORMULAS FOR SHEET METAL LAYOUT AND FABRICATION 6.13

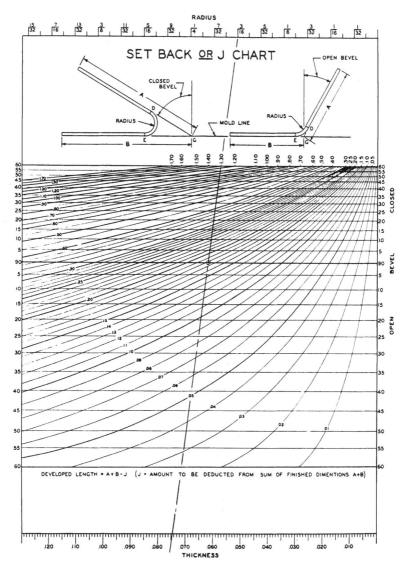

FIGURE 6.12 J chart for setback.

the equations were developed after extensive experimentation and practical working experience in the aerospace industry.

Bend Radii for Aluminum Alloys and Steel Sheets (Average). Figures 6.13 and 6.14 show average bend radii for various aluminum alloys and steel sheets. For other bend radii in different materials and gauges, see Table 6.1 for bend radii of different alloys, in terms of material thickness.

Material Gauge	Aluminum Alloy Designation		
	6061-T6 5052-H36 1100-H18	5052-H22 3003-H14	2024-T3
0.010	0.062	0.031	0.062
0.020	0.062	0.031	0.062
0.030	0.062	0.031	0.125
0.040	0.125	0.031	0.250
0.050	0.125	0.031	0.250
0.070	0.250	0.062	0.250
0.080	0.250	0.062	0.375
0.090	0.375	0.125	0.375
0.120	0.375	0.125	0.500
0.190	0.750	0.250	0.750
0.250	1.000	0.500	1.000

FIGURE 6.13 Bend radii for aluminum sheet metal.

6.2 SHEET METAL DEVELOPMENTS, TRANSITIONS, AND ANGLED CORNER FLANGE NOTCHING

The layout of sheet metal as required in development and transition parts is an important phase of sheet metal design and practice. The methods included here will prove useful in many design and working applications. These methods have application in ductwork, aerospace vehicles, automotive equipment, and other areas of product design and development requiring the use of transitions and developments.

Material Gauge	Steel Designation	
	AISI 1020	302-303-304S/S
0.010	0.031	0.031
0.020	0.031	0.031
0.030	0.031	0.031
0.040	0.031	0.031
0.050	0.031	0.031
0.060	0.031	0.062
0.070	0.031	0.062
0.080	0.031	0.062
0.090	0.062	0.062
0.120	0.062	0.125
0.190	0.125	0.250
0.250	0.125	0.250

FIGURE 6.14 Bend radii for steel sheets.

FORMULAS FOR SHEET METAL LAYOUT AND FABRICATION   6.15

TABLE 6.1 Minimum Bend Radii for Metals and Alloys in Multiples of Material Thickness, in

Material		0.015	0.031	0.063	0.093	0.125	0.188	0.250
Carbon steels								
SAE 1010		S	S	S	S	S	0.5	0.5
SAE 1020-1025		0.5	0.5	1.0	1.0	1.0	1.1	1.25
SAE 1070 & 1095		3.75	3.0	2.6	2.7	2.5	2.7	2.8
Alloy steels								
SAE 4130 & 8630		0.5	2.0	1.5	1.7	1.5	1.7	1.9
Stainless steels								
AISI 301, 302, 304 (A)		0.5 ————————————————					0.5	0.75
AISI 316 (A)		0.5 ————————————————					0.5	0.75
AISI 410, 430 (A)		1.0 ————————————————					1.0	1.25
AISI 301, 302, 304 (CR) ¼ H		0.5 ———————		0.5	1.0 ———————		1.0	1.25
AISI 316 ¼ H		1.0 ————————————————					1.0	1.25
AISI 301, 302, 304 ½ H		1.0					1.0	1.25
AISI 316 ½ H		2.0	2.0	3.0	2.0	2.0	2.0	2.5
AISI 301, 302, 304 H		2.0	2.0	1.5	1.5	1.5	1.5	1.5
Aluminum alloys								
1100	O	0	0	0	0	0	0	0
	H12	0	0	0	0	0	3.0	6.0
	H14	0	0	0	0	0	3.0	6.0
	H1	0	0	2.0	3.0	4.0	8.0	16.0
	H18	1.0	2.0	4.0	6.0	8.0	16.0	24.0
2014 & Alclad	O	0	0	0	0	0	3.0	6.0
	T6	2.0	4.0	8.0	15.0	20.0	36.0	64.0
2024 & Alclad	O	0	0	0	0	0	3.0	6.0
	T3	2.0	4.0	8.0	15.0	20.0	30.0	48.0
3003, 5005, 5357, 5457	O	0	0	0	0	0	0	0
	H12/H32	0	0	0	0	0	3.0	6.0
	H14/H34	0	0	0	1.0	2.0	4.0	8.0
	H16/H36	0	1.0	3.0	5.0	6.0	12.0	24.0
	H18/H38	1.0	2.0	5.0	9.0	12.0	24.0	40.0
5050, 5052, 5652	O	0	0	0	0	2.0	3.0	4.0
	H32	0	0	2.0	3.0	4.0	6.0	12.0
	H34	0	0	2.0	4.0	5.0	9.0	16.0
	H36	1.0	1.0	4.0	5.0	8.0	18.0	24.0
	H38	1.0	2.0	6.0	9.0	12.0	24.0	40.0
6061	O	0	0	0	0	2.0	3.0	4.0
	T6	1.0	2.0	4.0	6.0	9.0	18.0	28.0
7075 & Alclad	O	0	0	2.0	3.0	5.0	9.0	18.0
	T6	2.0	4.0	12.0	18.0	24.0	36.0	64.0
7178	O	0	0	2.0	3.0	5.0	9.0	18.0
	T6	2.0	4.0	12.0	21.0	28.0	42.0	80.0
Copper & alloys								
ETP 110	Soft	S	S	S	S	0.5	0.5	1.0
	Hard	S	1.0	1.5	2.0	2.0	2.0	2.0

(Continues)

TABLE 6.1 (*Continued*) Minimum Bend Radii for Metals and Alloys in Multiples of Material Thickness, in

Material		Thickness, in						
		0.015	0.031	0.063	0.093	0.125	0.188	0.250
Copper & alloys								
Alloy 210	¼ H	S	S	S	S	S	0.5	1.0
	½ H	S	S	S	S	S	1.0	1.5
	H	S	S	S	S	S	—	—
	EH	S	0.5	0.5	0.5	0.5	—	—
Alloy 260	¼ H	S	S	S	S	S	0.5	1.0
	½ H	S	S	S	0.3	0.3	—	—
	H	S	0.5	0.5	0.5	1.0	—	—
	EH	2.0	2.0	1.5	2.0	2.0	—	—
Alloy 353	¼ H	S	S	S	S	S	0.5	1.0
	½ H	S	0.5	0.5	0.7	0.3	—	—
	H	2.0	2.0	1.5	2.0	2.0	—	—
	EH	6.0	6.0	4.0	4.0	4.0	—	—
Magnesium sheet @ 70°F								
AZ31B-O (SB)		3.0						3.0
AZ31B-O		5.5						5.5
AZ31B-H24		8.0						8.0
HK31A-O		6.0						6.0
HK31A-H24		13.0						13.0
HM21A-T8		9.0						9.0
HM21A-T81		10.0						10.0
LA141A-O		3.0						3.0
ZE10A-O		5.5						5.5
ZE10A-H24		8.0						8.0
Titanium & alloys @ 70°F								
Pure (A)		3.0	3.0	3.0	3.5	3.5	3.5	3.5
Ti-8Mn (A)		4.0	4.0	4.0	4.0	5.0	5.0	5.0
Ti-5Al-2.5Sn (A)		5.5	5.5	5.5	5.5	6.0	6.0	6.0
Ti-6Al-4V (A)		4.5	4.5	4.5	5.0	5.0	5.0	5.0
Ti-6Al-4V (ST)		7.0						7.0
Ti-6Al-6V-2Sn (A)		4.0						4.0
Ti-13V-11Cr-3Al (A)		3.0	3.0	3.0	3.5	3.5	3.5	3.5
Ti-4Al-3Mo-1V (A)		3.5	3.5	3.5	4.0	4.0	4.0	4.0
Ti-4Al-3Mo-1V (ST)		5.5	5.5	5.5	6.0	6.0	6.0	6.0

Note: S = sharp bend; O = sharp bend; SB = special bending quality; A = annealed; ST = solution treated; H = hard; EH = extra hard. Magnesium sheet may be bent at temperatures to 800°F. Titanium may be bent at temperatures to 1000°F. On copper and alloys, direction of bending is at 90° to direction of rolling (bend radii must be increased 10 to 20 percent at 45° and 25 to 35 percent parallel to direction of rolling). The tabulated values of the minimum bend radii are given in multiples of the material thickness. The values of the bend radii should be tested on a test specimen prior to die design or production bending finished parts.

When sheet metal is to be formed into a curved section, it may be laid out, or *developed,* with reasonable accuracy by *triangulation* if it forms a simple curved surface without compound curves or curves in multiple directions. Sheet metal curved sections are found on many products, and if a straight edge can be placed flat against *elements* of the curved section, accurate layout or development is possible using the methods shown in this section.

FORMULAS FOR SHEET METAL LAYOUT AND FABRICATION   6.17

On double-curved surfaces such as are found on automobile and truck bodies and aircraft, forming dies are created from a full-scale model in order to duplicate these compound curved surfaces in sheet metal. The full-scale models used in aerospace vehicle manufacturing facilities are commonly called *mock-ups*, and the models used to transfer the compound curved surfaces are made by tool makers in the tooling department.

Skin Development (Outside Coverings).   Skin development on aerospace vehicles or other applications may be accomplished by triangulation when the surface is not double curved. Figure 6.15 presents a side view of the nose section of a simple aircraft. If we wish to develop the outer skin or sheet metal between stations 20.00 and 50.00, the general procedure is as follows: The *master lines* of the curves at stations 20.00 and 50.00 must be determined. In actual practice, the curves are developed by the master lines engineering group of the company, or you may know or develop your own curves. The procedure for layout of the flat pattern is as follows (see Fig. 6.16):

1. Divide curve *A* into a number of equally spaced points. Use the spline lengths (arc distances), *not* chordal distances.

2. Lay an accurate triangle tangent to one point on curve *A*, and by parallel action, transfer the edge of the triangle back to curve *B* and mark a point where the edge

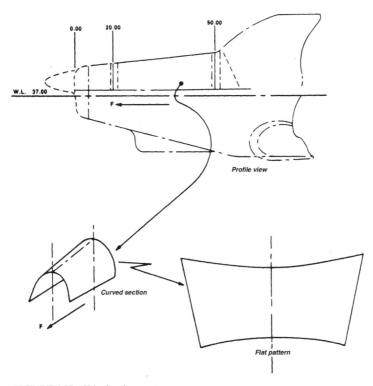

FIGURE 6.15   Skin development.

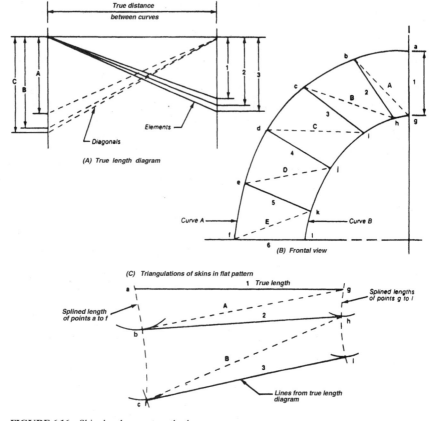

FIGURE 6.16 Skin development method.

of the triangle is tangent to curve B (e.g., point b on curve A back to point h on curve B; see Fig. 6.16b). Then parallel transfer all points on curve A back to curve B and label all points for identification. Draw the element lines and diagonals on the frontal view, that is, 1A, 2B, 3C, etc.

3. Construct a true-length diagram as shown in Fig. 6.16a, where all the element and diagonal true lengths can be found (elements are 1, 2, 3, 4, etc.: diagonals are A, B, C, D, etc.). The true distance between the two curves is 30.00; that is, 50.00 − 20.00, from Fig. 6.15.

4. Transfer the element and diagonal true-length lines to the triangulation flat-pattern layout as shown in Fig. 6.16c. The triangulated flat pattern is completed by transferring all elements and diagonals to the flat-pattern layout.

Canted-Station Skin Development (Bulkheads at an Angle to Axis). When the planes of the curves *A* and *B* (Fig. 6.17) are *not* perpendicular to the axis of the curved section, layout procedures to determine the true lengths of the element and

FORMULAS FOR SHEET METAL LAYOUT AND FABRICATION 6.19

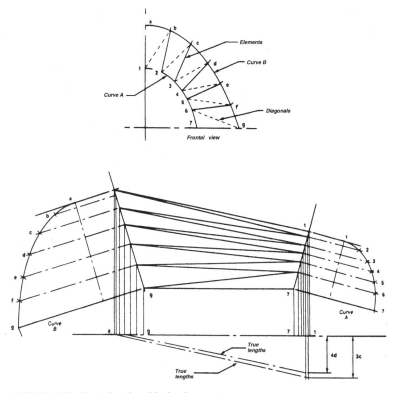

FIGURE 6.17 Canted-station skin development.

diagonal lines are as shown in Fig. 6.17. The remainder of the procedure is as explained in Fig. 6.16 to develop the triangulated flat pattern.

In aerospace terminology, the locations of points on the craft are determined by *station, waterline,* and *buttline.* These terms are defined as follows:

Station: The numbered locations from the front to the rear of the craft.

Waterline: The vertical locations from the lowest point to the highest point of the craft.

Buttline: The lateral locations from the centerline of the axis of the craft to the right and to the left of the axis of the craft. There are right buttlines and left buttlines.

With these three axes, any exact point on the craft may be described or dimensioned.

Developing Flat Patterns for Multiple Bends. Developing flat patterns can be done by bend deduction or setback. Figure 6.18 shows a type of sheet metal part that may be bent on a press brake. The flat-pattern part is bent on the brake, with the center of bend line (CBL) held on the bending die centerline. The machine's back gauge

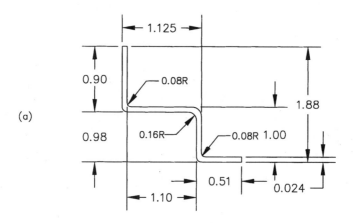

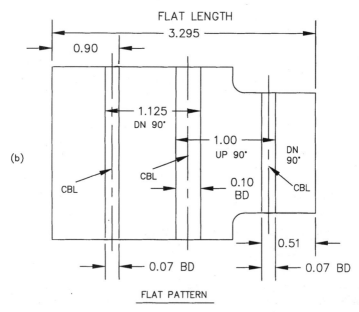

FIGURE 6.18 Flat pattern development.

is set by the operator in order to form the part. If you study the figure closely, you can see how the dimensions progress: The bend deduction is drawn in, and the next dimension is taken from the end of the first bend deduction. The next dimension is then measured, the bend deduction is drawn in for that bend, and then the next dimension is taken from the end of the second bend deduction, etc. Note that the second bend deduction is larger because of the larger radius of the second bend (0.16R).

FORMULAS FOR SHEET METAL LAYOUT AND FABRICATION   6.21

Stiffening Sheet Metal Parts. On many sheet metal parts that have large areas, stiffening can be achieved by creasing the metal in an X configuration by means of brake bending. On certain parts where great stiffness and rigidity are required, a method called *beading* is employed. The beading is carried out at the same time as the part is being hydropressed, Marformed, or hard-die formed. See Sec. 6.5 for data on beading sheet metal parts, and other tooling requirements for sheet metal.

Another method for stiffening the edge of a long sheet metal part is to hem or *Dutch bend* the edge. In aerospace and automotive sheet metal parts, flanged *lightening holes* are used. The lightening hole not only makes the part lighter in weight but also more rigid. This method is used commonly in wing ribs, airframes, and gussets or brackets. The lightening hole need not be circular but can take any convenient shape as required by the application.

Typical Transitions and Developments. The following transitions and developments are the most common types, and learning or using them for reference will prove helpful in many industrial applications. Using the principles shown will enable you to apply these to many different variations or geometric forms.

Development of a Truncated Right Pyramid. Refer to Fig. 6.19. Draw the projections of the pyramid that show (1) a normal view of the base or right section and (2) a normal view of the axis. Lay out the pattern for the pyramid and then superimpose the pattern on the truncation.

Since this is a portion of a right regular pyramid, the lateral edges are all of equal length. The lateral edges OA and OD are parallel to the frontal plane and consequently show in their true length on the front view. With the center at O_1, taken at any convenient place, and a radius $O_F A_F$, draw an arc that is the stretchout of the

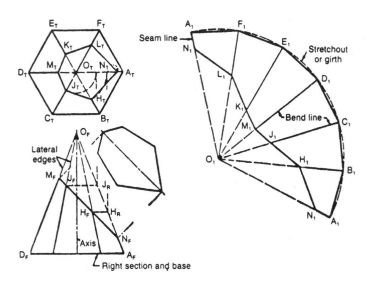

FIGURE 6.19 Development of a truncated right pyramid.

pattern. On it, step off the six equal sides of the hexagonal base obtained from the top view, and connect these points successively with each other and with the vertex O_1, thus forming the pattern for the pyramid.

The intersection of the cutting plane and lateral surfaces is developed by laying off the true length of the intercept of each lateral edge on the corresponding line of the development. The true length of each of these intercepts, such as OH, OJ, etc., is found by rotating it about the axis of the pyramid until it coincides with O_FA_F as previously explained. The path of any point, such as H, will be projected on the front view as a horizontal line. To obtain the development of the entire surface of the truncated pyramid, attach the base; also find the true size of the cut face, and attach it on a common line.

Development of an Oblique Pyramid. Refer to Fig. 6.20. Since the lateral edges are unequal in length, the true length of each must be found separately by rotating it parallel to the frontal plane. With O_1 taken at any convenient place, lay off the seam line O_1A_1 equal to O_FA_R. With A_1 as center and radius O_1B_1 equal to O_FB_R, describe a second arc intersecting the first in vertex B_1. Connect the vertices O_1, A_1, and B_1, thus forming the pattern for the lateral surface OAB. Similarly, lay out the pattern for the remaining three lateral surfaces, joining them on their common edges. The stretchout is equal to the summation of the base edges. If the complete development is required, attach the base on a common line.

Development of a Truncated Right Cylinder. Refer to Fig. 6.21. The development of a cylinder is similar to the development of a prism. Draw two projections of the cylinder:

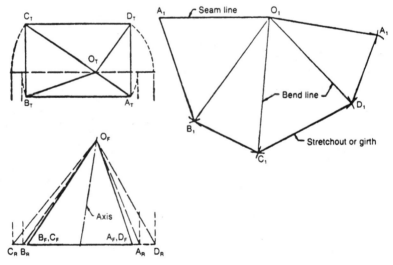

FIGURE 6.20 Development of an oblique pyramid.

FORMULAS FOR SHEET METAL LAYOUT AND FABRICATION   6.23

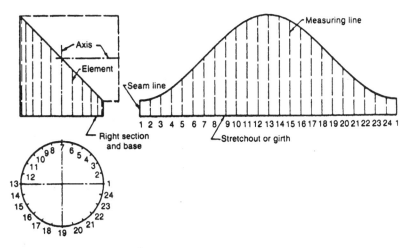

FIGURE 6.21   Development of a truncated right cylinder.

1. A normal view of a right section
2. A normal view of the elements

In rolling the cylinder out on a tangent plane, the base or right section, being perpendicular to the axis, will develop into a straight line. For convenience in drawing, divide the normal view of the base, shown here in the bottom view, into a number of equal parts by points that represent elements. These divisions should be spaced so that the chordal distances approximate the arc closely enough to make the stretchout practically equal to the periphery of the base or right section.

Project these elements to the front view. Draw the stretchout and measuring lines, the cylinder now being treated as a many-sided prism. Transfer the lengths of the elements in order, either by projection or by using dividers, and join the points thus found by a smooth curve. Sketch the curve in very lightly, freehand, before fitting the French curve or ship's curve to it. This development might be the pattern for one-half of a two-piece elbow.

Three-piece, four-piece, and five-piece elbows may be drawn similarly, as illustrated in Fig. 6.22. Since the base is symmetrical, only one-half of it need be drawn. In these cases, the intermediate pieces such as *B, C,* and *D* are developed on a stretchout line formed by laying off the perimeter of a right section. If the right section is taken through the middle of the piece, the stretchout line becomes the center of the development. Evidently, any elbow could be cut from a single sheet without waste if the seams were made alternately on the long and short sides.

Development of a Truncated Right Circular Cone.   Refer to Fig. 6.23. Draw the projection of the cone that will show (1) a normal view of the base or right section and (2) a normal view of the axis. First, develop the surface of the complete cone and then superimpose the pattern for the truncation.

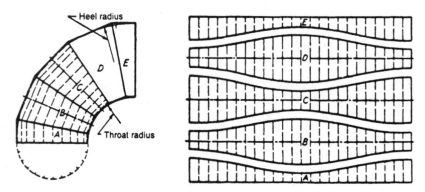

FIGURE 6.22 Development of a five-piece elbow.

Divide the top view of the base into a sufficient number of equal parts that the sum of the resulting chordal distances will closely approximate the periphery of the base. Project these points to the front view, and draw front views of the elements through them. With center A_1 and a radius equal to the slant height $A_F I_F$, which is the true length of all the elements, draw an arc, which is the stretchout. Lay off on it the chordal divisions of the base, obtained from the top view. Connect these points 2, 3, 4, 5, etc. with A_1, thus forming the pattern for the cone.

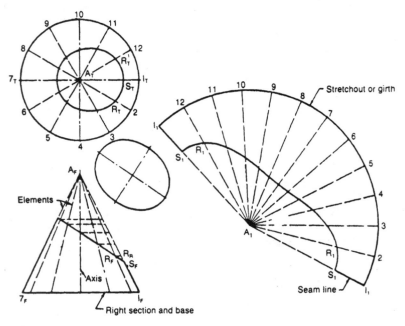

FIGURE 6.23 Development of a truncated circular cone.

Find the true length of each element from vertex to cutting plane by rotating it to coincide with the contour element A_1, and lay off this distance on the corresponding line of the development. Draw a smooth curve through these points. The pattern for the cut surface is obtained from the auxiliary view.

Triangulation. Nondevelopable surfaces are developed approximately by assuming them to be made of narrow sections of developable surfaces. The most common and best method for approximate development is *triangulation;* that is, the surface is assumed to be made up of a large number of triangular strips or plane triangles with very short bases. This method is used for all warped surfaces as well as for oblique cones. Oblique cones are single-curved surfaces that are capable of true theoretical development, but they can be developed much more easily and accurately by triangulation.

Development of an Oblique Cone. Refer to Fig. 6.24. An oblique cone differs from a cone of revolution in that the elements are all of different lengths. The development of a right circular cone is made up of a number of equal triangles meeting at the vertex whose sides are elements and whose bases are the chords of short arcs of the base of the cone. In the oblique cone, each triangle must be found separately.

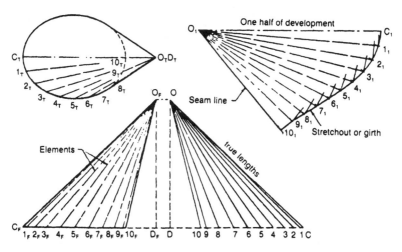

FIGURE 6.24 Development of an oblique cone.

Draw two views of the cone showing (1) a normal view of the base and (2) a normal view of the altitude. Divide the true size of the base, shown here in the top view, into a number of equal parts such that the sum of the chordal distances will closely approximate the length of the base curve. Project these points to the front view of the base. Through these points and the vertex, draw the elements in each view.

Since the cone is symmetrical about a frontal plane through the vertex, the elements are shown only on the front half of it. Also, only one-half of the development is drawn. With the seam on the shortest element, the element OC will be the centerline of the development and may be drawn directly at O_1C_1, since its true length is given by O_FC_F.

Find the true length of the elements by rotating them until they are parallel to the frontal plane or by constructing a *true-length diagram*. The true length of any element will be the hypotenuse of a triangle with one leg the length of the projected element, as seen in the top view, and the other leg equal to the altitude of the cone. Thus, to make the diagram, draw the leg OD coinciding with or parallel to O_FD_F. At D and perpendicular to OD, draw the other leg, and lay off on it the lengths $D1, D2$, etc. equal to D_T1_T, D_T2_T, etc., respectively. Distances from point O to points on the base of the diagram are the true lengths of the elements.

Construct the pattern for the front half of the cone as follows. With O_1 as the center and radius $O1$, draw an arc. With C_1 as center and the radius C_T1_T, draw a second arc intersecting the first at 1_1. Then O_11_1 will be the developed position of the element $O1$. With 1_1 as the center and radius 1_T2_T, draw an arc intersecting a second arc with O_1 as center and radius $O2$, thus locating 2_1. Continue this procedure until all the elements have been transferred to the development. Connect the points $C_1, 1_1, 2_1$, etc. with a smooth curve, the stretchout line, to complete the development.

Conical Connection Between Two Cylindrical Pipes. Refer to Fig. 6.24. The method used in drawing the pattern is the application of the development of an oblique cone. One-half the elliptical base is shown in true size in an auxiliary view (here attached to the front view). Find the true size of the base from its major and minor axes; divide it into a number of equal parts so that the sum of these chordal distances closely approximates the periphery of the curve. Project these points to the front and top views. Draw the elements in each view through these points, and find the vertex O by extending the contour elements until they intersect.

The true length of each element is found by using the vertical distance between its ends as the vertical leg of the diagram and its horizontal projection as the other leg. As each true length from vertex to base is found, project the upper end of the intercept horizontally across from the front view to the true length of the corresponding element to find the true length of the intercept. The development is drawn by laying out each triangle in turn, from vertex to base, as in Fig. 6.25, starting on the centerline O_1C_1, and then measuring on each element its intercept length. Draw smooth curves through these points to complete the pattern.

Development of Transition Pieces. Refer to Figs. 6.26 and 6.27. Transitions are used to connect pipes or openings of different shapes or cross sections. Figure 6.26, showing a transition piece for connecting a round pipe and a rectangular pipe, is typical. These pieces are always developed by triangulation. The piece shown in Fig. 6.26 is, evidently, made up of four triangular planes whose bases are the sides of the rectangle and four parts of oblique cones whose common bases are arcs of the circle and whose vertices are at the corners of the rectangle. To develop the piece, make a true-length diagram as shown in Fig. 6.24. The true length of $O1$ being found, all the sides of triangle A will be known. Attach the developments of cones B and B^1, then those of triangle C and C^1, and so on.

FORMULAS FOR SHEET METAL LAYOUT AND FABRICATION 6.27

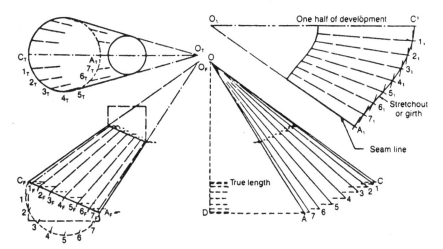

FIGURE 6.25 Development of a conical connection between two cylinders.

Figure 6.27 is another transition piece joining a rectangle to a circular pipe whose axes are not parallel. By using a partial right-side view of the round opening, the divisions of the bases of the oblique cones can be found. (Since the object is symmetrical, only one-half the opening need be divided.) The true lengths of the elements are obtained as shown in Fig. 6.26.

Triangulation of Warped Surfaces. The approximate development of a warped surface is made by dividing it into a number of narrow quadrilaterals and then split-

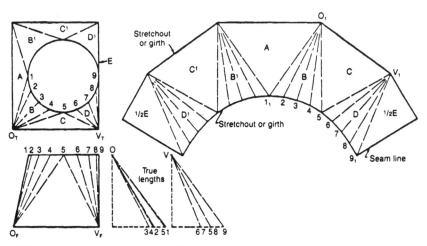

FIGURE 6.26 Development of a transition piece.

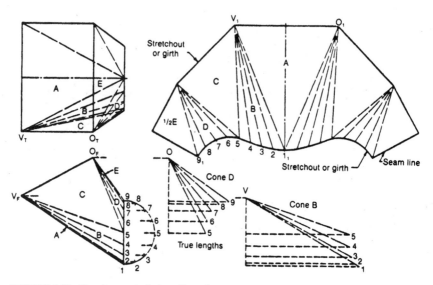

FIGURE 6.27 Development of a transition piece.

ting each of these into two triangles by a diagonal line, which is assumed to be a straight line, although it is really a curve. Figure 6.28 shows a warped transition piece that connects on ovular (upper) pipe with a right-circular cylindrical pipe (lower). Find the true size of one-half the elliptical base by rotating it until horizontal about an axis through 1, when its true shape will be seen.

Sheet Metal Angled Corner Flange Notching: Flat-Pattern Development. Sheet metal parts sometimes have angled flanges that must be bent up for an exact angular fit. Figure 6.29 shows a typical sheet metal part with 45° bent-up flanges. In order to lay out the corner notch angle for this type of part, you may use PC programs such as AutoCad to find the correct dimensions and angular cut at the corners, or you may calculate the corner angular cut by using trigonometry. To trigonometrically calculate the corner angular notch, proceed as follows:

From Fig. 6.29, sketch the flat-pattern edges and true lengths as shown in Fig. 6.30, forming a triangle ABC which may now be solved by first using the law of cosines to find side b, and then the law of sines to determine the corner half-notch angle, angle C.

The triangle ABC begins with known dimensions: $a = 4$, angle $B = 45°$, and $c = 0.828$. That is a triangle where you know two sides and the included angle B. You will need to first find side b, using the law of cosines as follows:

$$b^2 = a^2 + c^2 - 2ac \cos B$$

$$b^2 = (4)^2 + (0.828)^2 - 2 \cdot 4 \cdot 0.828 \cdot 0.707 \quad \text{(by the law of cosines)}$$

$$b^2 = 12.00242$$

$$b = \sqrt{12} = 3.464$$

FORMULAS FOR SHEET METAL LAYOUT AND FABRICATION 6.29

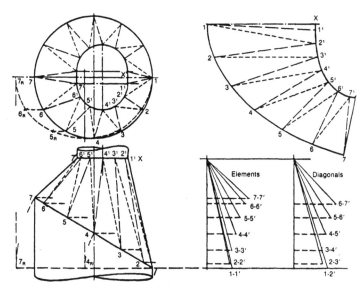

FIGURE 6.28 Development of a warped transition piece.

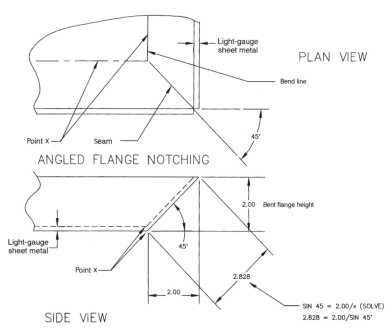

SIN 45 = 2.00/x (SOLVE)
2.828 = 2.00/SIN 45°

FIGURE 6.29 Angled flanges.

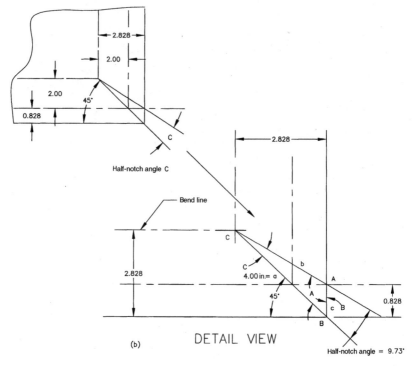

FIGURE 6.30 Layout of angled flange notching.

Then, find angle C using the law of sines: $\sin B/b = \sin C/c$.

$$\frac{\sin B}{3.464} = \frac{\sin C}{0.828}$$

$$\frac{\sin 45}{3.464} = \frac{\sin C}{0.828}$$

$3.464 \sin C = 0.707 \cdot 0.828$ (by the law of sines)

$$\sin C = \frac{0.5854}{3.464}$$

$\sin C = 0.16899$

$\arcsin 0.16899 = 9.729$

$\therefore C = 9°44'$

Therefore, the notch angle $= 2 \times 9°44' = 19°28'$.

FORMULAS FOR SHEET METAL LAYOUT AND FABRICATION   6.31

This procedure may be used for determining the notch angle for flanges bent on any angle.

The angle given previously as 19°28′ is valid for any flange length, as long as the bent-up angle is 45°. This notch angle will increase as the bent-up flanges approach 90°, until the angle of notch becomes 90° for a bent-up angle of 90°.

NOTE. The triangle ABC shown in this example is actually the overlap angle of the metal flanges as they become bent up 45°, which must be removed as the corner notch. On thicker sheet metal, such as 16 through 7 gauge, you should do measurements and the calculations from the inside mold line (IML) of the flat-pattern sheet metal. Also, the flange height, shown as 2 in Fig. 6.29, could have been 1 in, or any other dimension, in order to do the calculations. Thus, the corner notch angle is a *constant* angle for every given bent-up angle; i.e., the angular notch for all 45° bent-up flanges is always 19°28′, and will always be a *different* constant angular notch for every *different* bent-up flange angle.

Figure 6.31 shows an AutoCad scale drawing confirming the calculations given for Figs. 6.29 and 6.30.

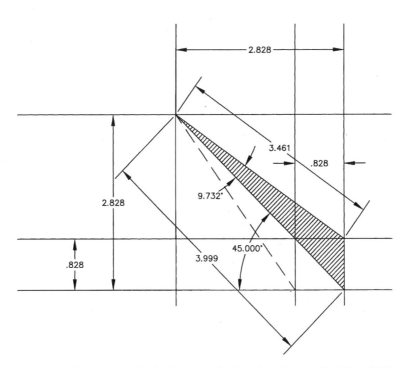

FIGURE 6.31   An AutoCad scaled layout confirming calculations for Figs. 6.29 and 6.30. Note that the shaded area is the half-notch cutout.

6.3 PUNCHING AND BLANKING PRESSURES AND LOADS

Force Required for Punching or Blanking. The simple equation for calculating the punching or blanking force P in pounds for a given material and thickness is given as

$$P = SLt \quad \text{For any shape or aperture}$$

$$P = S\pi Dt \quad \text{For round holes}$$

where P = force required to punch or blank, lbf
S = shear strength of material, psi (see Fig. 6.32)
L = sheared length, in
D = diameter of hole, in
t = thickness of material, in

Stripping Forces. Stripping forces vary from 2.5 to 20 percent of the punching or blanking forces. A frequently used equation for determining the stripping forces is

$$F_s = 3500Lt$$

where F_s = stripping force, lbf
L = perimeter of cut (sheared length), in
t = thickness of material, in

NOTE. This equation is approximate and may not be suitable for all conditions of punching and blanking due to the many variables encountered in this type of metalworking.

6.4 SHEAR STRENGTHS OF VARIOUS MATERIALS

The shear strength (in pounds per square inch) of the material to be punched or blanked is required in order to calculate the force required to punch or blank any particular part. Figure 6.32 lists the average shear strengths of various materials, both metallic and nonmetallic. If you require the shear strength of a material that is not listed in Fig. 6.32, an approximation of the shear strength may be made as follows (for relatively ductile materials only): Go to a handbook on materials and their uses, and find the *ultimate tensile strength* of the given material. Take 45 to 55 percent of this value as the approximate shear strength. For example, if the ultimate tensile strength of the given material is 75,000 psi,

Shear strength = $0.45 \times 75,000 = 30,750$ psi approximately (low value)

$= 0.55 \times 75,000 = 41,250$ psi approximately (high value)

FORMULAS FOR SHEET METAL LAYOUT AND FABRICATION 6.33

Material	Shear Strength, psi
Carbon Steels:	
SAE 1010 HR	21,500
SAE 1020 HR	32,000
SAE 1045 QT	55,000
SAE 1045 A	44,000
SAE 1095 QT	90,000
SAE 1095 A	63,000
SAE 1117 HR	32,000
Alloy Steels:	
SAE 4130 N	43,500
SAE 4130 T (150,000)	90,000
SAE 4140 N	66,500
SAE 3120 HT-D (800°F)	95,000
SAE 3140 HT-D (800°F)	130,000
SAE 3250 HT-D (800°F)	165,000
Stainless Steels:	
AISI 201	52,000
AISI 301	50,000
AISI 302	41,000
AISI 304	38,500
AISI 310	42,750
AISI 316	38,250
AISI 321	38,250
Cold rolled S/S strip (full hard)	
AISI 300 Series	112,000
Stainless Steels: Annealed	
AISI 410	33,750
AISI 416	33,750
AISI 440C	49,500
AISI 430	33,750
Monel Metal:	
70,000 UTS	42,900
110,000 UTS	65,500
K Monel:	
155,500 UTS	98,500
Nickel:	
68,000 UTS	52,300
121,000 UTS	75,300
Inconel Alloys:	
80,000 UTS	59,000
100,000 UTS	66,000
150,000 UTS	80,000
175,000 UTS	87,000

FIGURE 6.32 Shear Strengths of Metallic and Nonmetallic Materials—psi

Copper and Alloys:	
CA 110 (ETP 110)	22,000-28,000
CA 210 (Guilding)	26,000-37,000
CA 220 (Bronze)	28,000-38,000
CA 230 (Red brass)	31,000-42,000
CA 260 (Cartridge brass)	33,000-44,000
CA 268 (Yellow brass)	33,000-43,000
Beryllium copper: Strip & sheet	
C 17200 (25)	34,200-54,000
C 17000 (165)	34,200-94,500
C 17510 (3)	24,750-67,500
C 17500 (10)	24,750-67,500
C 17410 (174) HT	58,500
Beryllium Nickel:	
UNS-N033 HT	123,750
Aluminum and Alloys:	
1100-O	9,000
1100-H18	13,000
2014-O	18,000
2014-T4, T451	38,000
2014-T6, T651	42,000
2024-O	18,000
2024-T3, T4, T351	41,000
3003-O	11,000
3003-H14	14,000
3003-H18	16,000
5052-O	18,000
5052-H32, H38	60,000-77,000
6061-O	12,000
6061-T4, T451	24,000
6061-T6, T651	30,000
7075-O	22,000
7075-T6, T651	48,000
7178-O	23,000
7178-T6, T651	53,300
Magnesium Alloys:	
Soft (annealed)	19,000
Hardened	28,500 max.
Titanium & Alloys:	
Pure	27,000-49,500
Typical alloys	45,000-77,000
Nonmetallics:	
Polyester-glass (GPO-1, 2 & 3)	12,000-17,000
Polycarbonate (Lexan)	6,000-10,000
Cycolac	4,400-7,400
ABS (Acrylonitrile Butadene Styrene)	1,500-4,000
Acetal (Delrin)	3,000-6,000
Acetate (Cellulose)	2,000-4,000
Epoxy-glass	4,000-10,000
Nylon	3,000-12,000

FIGURE 6.32 *(Continued)* Shear Strengths of Metallic and Nonmetallic Materials—psi

FORMULAS FOR SHEET METAL LAYOUT AND FABRICATION   6.35

Phenolic resins (cloth)	26,000 (Hot-blanked)
Paper	3,500-6,400
Mica	10,000
Teflon, rigid (TFE)	1,500-3,000
Hard rubber	20,000
Polystyrene	10,000 max.
Asbestos board	5,000

Notes: For metallic materials- when the tabulated shear values are given in ranges, the shear values run from the annealed or soft condition to the hardest condition. Interpolate intermediate values between ranges. For nonmetallic materials- the shear value ranges are given from soft to hard grades or glass-filled grades.

A = annealed; HR = hot-rolled; N = normalized; T = tempered; HT-D = heat-treated and drawn (tempered); QT = quenched and tempered; UTS = ultimate tensile strength; HT = heat-treated.

FIGURE 6.32 *(Continued)* Shear Strengths of Metallic and Nonmetallic Materials—psi

Manufacturers' Standard Gauges for Steel Sheets. The decimal equivalents of the American standard manufacturers' gauges for steel sheets are shown in Figs. 6.4 and 6.5. Sheet steels in the United States are purchased to these gauge equivalents, and tools and dies are designed for this standard gauging system.

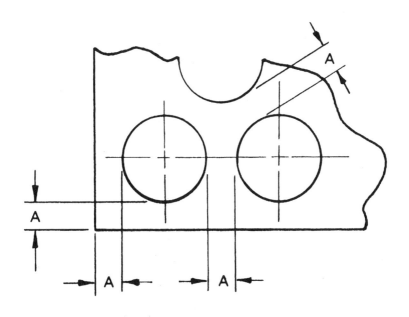

Material   **Minimum Distance**

Up to 0.062"   A = 0.12 Min.
Over 0.062"   A = 2 x Metal Thickness
(a)

FIGURE 6.33 Punching requirements.

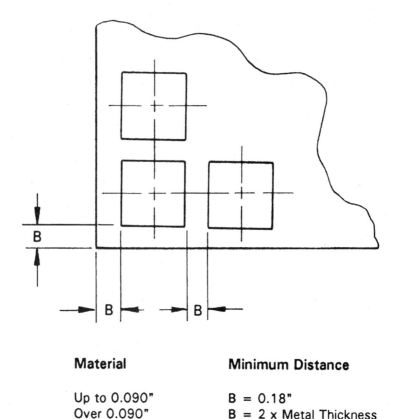

Material   **Minimum Distance**

Up to 0.090"   B = 0.18"
Over 0.090"   B = 2 × Metal Thickness

(b)

FIGURE 6.33 *(Continued)* Punching requirements.

6.5 TOOLING REQUIREMENTS FOR SHEET METAL PARTS—LIMITATIONS

Minimum distances for hole spacings and edge distances for punched holes in sheet parts are shown in Figs. 6.33a and b. Following these guidelines will prevent buckling or tearing of the sheet metal.

Corner relief notches for areas where a bent flange is required is shown in Fig. 6.34a. The minimum edge distance for angled flange chamfer height is shown in Fig. 6.34b. The X dimension in Fig. 6.34b is determined by the height from the center of the bend radius ($2 \times t$). If the inside bend radius is 0.25 in, and the material thickness is 0.125 in, the dimension X would be:

$$0.25 + 0.125 + (2 \times 0.125) = 0.625 \text{ in}$$

or $X = 2t + R$, per the figure.

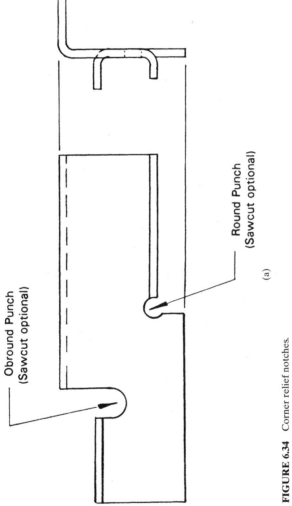

FIGURE 6.34 Corner relief notches.

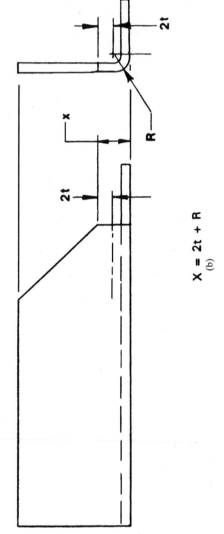

FIGURE 6.34 (*Continued*) Corner relief notches.

FORMULAS FOR SHEET METAL LAYOUT AND FABRICATION

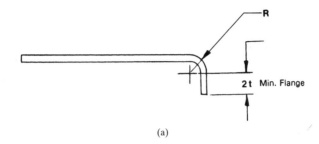

(a)

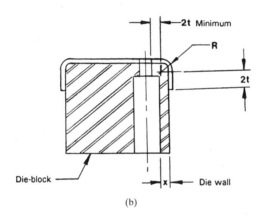

(b)

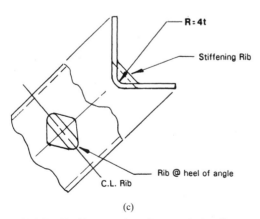

(c)

FIGURE 6.35 Sheet metal requirements for bending.

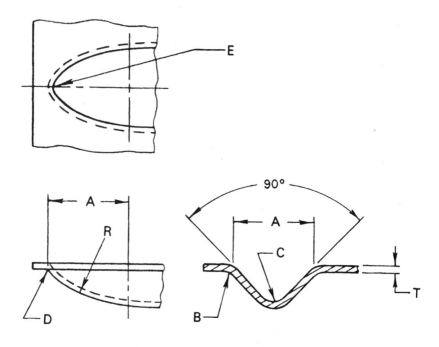

A	B (Radius)	C (Radius)	D (Radius)	E (Radius)
0.25	2T	2T	4T	T
0.38	2T	2T	4T	T
0.50	2T	2T	4T	2T
0.62	4T	4T	4T	2T
0.75	5T	5T	4T	3T
1.00	5T	5T	4T	3T

(a)

FIGURE 6.36 Stiffening beads in sheet metal.

Minimum flanges on bent sheet metal parts are shown in Fig. 6.35a. Flanges' and holes' minimum dimensions are shown in Fig. 6.35b. Bending dies are usually employed to achieve these dimensions, although on a press brake, *bottoming* dies may be used if the gauges are not too heavy.

Stiffening ribs placed in the heel of sheet metal angles should maintain the dimensions shown in Fig. 6.35c.

FORMULAS FOR SHEET METAL LAYOUT AND FABRICATION

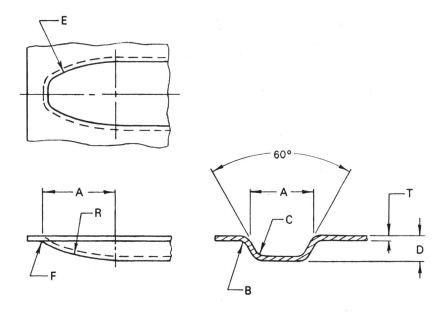

A	B (Radius)	C (Radius)	D	E (Radius)	F (Radius)
1.00	3T	2T	0.25	4T	4T
1.50	3T	2T	0.31	4T	4T

(b)

FIGURE 6.36 (*Continued*) Stiffening beads in sheet metal.

Stiffening beads placed in the webs of sheet metal parts for stiffness should be controlled by the dimensions shown in Figs. 6.36a and b. The dimensions shown in these figures determine the allowable depth of the bead, which depends on the thickness (gauge) of the material.

CHAPTER 7
GEAR AND SPROCKET CALCULATIONS

7.1 INVOLUTE FUNCTION CALCULATIONS

Involute functions are used in some of the equations required to perform involute gear design. These functional values of the involute curve are easily calculated with the aid of the pocket calculator. Refer to the following text for the procedure required to calculate the involute function.

The Involute Function: inv ϕ = tan ϕ - arc ϕ. The involute function is widely used in gear calculations. The angle ϕ for which involute tables are tabulated is the slope of the involute with respect to a radius vector R (see Fig. 7.1).

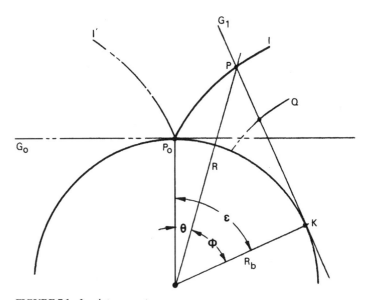

FIGURE 7.1 Involute geometry.

Involute Geometry (See Fig. 7.1). The *involute* of a circle is defined as the curve traced by a point on a straight line which rolls without slipping on the circle. It is also described as the curve generated by a point on a nonstretching string as it is unwound from a circle. The circle is called the *base circle* of the involute. A single involute curve has two branches of opposite hand, meeting at a point on the base circle, where the radius of curvature is zero. All involutes of the same base circle are congruent and parallel, while involutes of different base circles are geometrically similar.

Figure 7.1 shows the elements of involute geometry. The generating line was originally in position G_0, tangent to the base circle at P_0. The line then rolled about the base circle through the roll angle ε to position G_1, where it is tangent to the base circle at K. The point P_0 on the generating line has moved to P, generating the involute curve I. Another point on the generating line, such as Q, generates another involute curve which is congruent and parallel to curve I.

Since the generating line is always normal to the involute, the angle φ is the slope of the involute with respect to the radius vector R. The polar angle θ together with R constitute the coordinates of the involute curve. The parametric polar equations of the involute are

$$R = R_b \sec \phi$$

$$\theta = \tan \phi - \tilde{\phi}$$

The quantity $(\tan \phi - \tilde{\phi})$ is called the *involute function* of $\tilde{\phi}$.

NOTE. The roll angle ε in radians is equal to tan φ.

Calculating the Involute Function (inv φ = tan φ − arc φ). Find the involute function for 20.00°.

$$\text{inv } \phi = \tan \phi - \text{arc } \phi$$

where   tan φ = natural tangent of the given angle
arc φ = numerical value, in radians, of the given angle

Therefore,

$$\text{inv } \phi = \tan 20° - 20° \text{ converted to radians}$$

$$\text{inv } \phi = 0.3639702 - (20 \times 0.0174533)$$

NOTE. 1° = 0.0174533 rad.

$$\text{inv } \phi = 0.3639702 - 0.3490659$$

$$\text{inv } 20° = 0.0149043$$

The involute function for 20° is 0.0149043 (accurate to 7 decimal places).

Using the procedure shown here, it becomes *obvious* that a table of involute functions is not required for gearing calculation procedures. It is also safer to calculate your own involute functions because handbook tables may contain typographical errors.

GEAR AND SPROCKET CALCULATIONS 7.3

EXAMPLE. To plot an involute curve for a base circle of 3.500-in diameter, proceed as follows. Refer to Fig. 7.2 and the preceding equations for the x and y coordinates. The solution for angle $\theta = 60°$ will be calculated longhand, and then the MathCad program will be used to calculate all coordinates from 0° to 120°, by using *range variables* in nine 15° increments.

NOTE. Angle θ must be given in radians; 1 rad = $\pi/180° = 0.0174532$; $2\pi R = 360°$.

$x = r \cos \theta + r\theta \sin \theta$

$x = 2.750 \cos \left[60\left(\dfrac{\pi}{180}\right) \right] + r\left[60\left(\dfrac{\pi}{180}\right) \right] \sin \left[60\left(\dfrac{\pi}{180}\right) \right]$

$x = 2.750 \cos (1.04719755) + [2.750(1.04719755) \sin (1.04719755)]$

$x = 2.750(0.5000000) + [2.750(1.04719755)(0.8660254)]$

$x = 1.37500 + 2.493974$

$x = 3.868974$

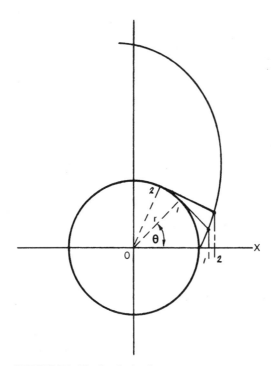

FIGURE 7.2 Plotting the involute curve.

$y = r \sin \theta - r\theta \cos \theta$

$y = 2.750 \sin\left[60\left(\dfrac{\pi}{180}\right)\right] - \left\{2.750\left[60\left(\dfrac{\pi}{180}\right)\right] \cos\left[60\left(\dfrac{\pi}{180}\right)\right]\right\}$

$y = 2.750(0.8660254) - [2.750(1.04719755)(0.500000)]$

$y = 2.3815699 - 1.4398966$

$y = 0.9416733$

Therefore, the x ordinate is 3.868974, and the y ordinate is 0.9416733.

The MathCad 8 calculation sheet seen in Fig. 7.3 will give the complete set of coordinates for the x and y axes, to describe the involute curve from $\theta = 0$ to 120°. The coordinates just calculated check with the MathCad calculation sheet for 60°.

Plotting the Involute Curve (See Fig. 7.2). The x and y coordinates of the points on an involute curve may be calculated from

$x = r \cos \theta + r\theta \sin \theta$

$y = r \sin \theta - r\theta \cos \theta$

Calculating the Inverse Involute Function. Calculating the involute function for a given angle is an easy proposition, as shown in the previous calculations. But the problem of calculating the angle ϕ for a given involute function θ is difficult, to say the least. In certain gearing and measurement equations involving involute functions, it is sometimes required to find the angle ϕ for a given involute function θ. In the past, this was done by calculating an extensive table of involute functions from an extensive number of angles, in small increments of minutes.

But since you don't know the angle for all involute functions, this can be a very tedious task. The author has developed a mathematical procedure for calculating the angle for *any* given involute function. The procedure involves the infinite series for the sine and cosine, using the MathCad 8 program. A self-explanatory example is shown in Fig. 7.4, where the unknown angle ϕ is solved for a given arbitrary value of the involute function θ. This procedure is valid for all angles ϕ from any involute function value θ.

MathCad 8 solves for all roots for angle ϕ in radians, and only one of the many roots representing the involute function is applicable, as shown in Fig. 7.4. This procedure is then aptly termed finding the *inverse involute function.*

7.2 GEARING FORMULAS—SPUR, HELICAL, MITER/BEVEL, AND WORM GEARS

The standard definitions for spur gear terms are shown in Fig. 7.5.

Equivalent diametral pitch (DP), circular pitch (CP), and module values are shown in Fig. 7.6. DP and CP are U.S. customary units and module values are SI units.

GEAR AND SPROCKET CALCULATIONS 7.5

Range variables

Radius of base circle
$r := 2.750$

$$\theta := 0 \cdot \left(\frac{\pi}{180}\right), 15 \cdot \left(\frac{\pi}{180}\right) .. 120 \cdot \left(\frac{\pi}{180}\right)$$

Range variables in 15-degree increments, expressed in radians from 0 to 120 degrees.

x ordinates	y ordinates
$r \cdot \cos(\theta) + r \cdot \theta \cdot \sin(\theta) =$	$r \cdot \sin(\theta) - r \cdot \theta \cdot \cos(\theta) =$
2.75	0
2.84263236	0.0163357
3.10151818	0.12801294
3.47178466	0.41730264
3.86897413	0.94167323
4.18883574	1.72461434
4.3196899	2.75
4.15616433	3.96065037
3.61294825	5.26136313

$$r \cdot \cos\left[45 \cdot \left(\frac{\pi}{180}\right)\right] + r \cdot \left[45 \cdot \left(\frac{\pi}{180}\right)\right] \cdot \sin\left[45 \cdot \left(\frac{\pi}{180}\right)\right] = 3.47178466 \quad \text{x ordinate (calculated for 45 degrees)}$$

$$r \cdot \sin\left[45 \cdot \left(\frac{\pi}{180}\right)\right] - r \cdot \left[45 \cdot \left(\frac{\pi}{180}\right)\right] \cdot \cos\left[45 \cdot \left(\frac{\pi}{180}\right)\right] = 0.41730264 \quad \text{y ordinate (calculated for 45 degrees)}$$

Note: The angle θ in the above calculations is expressed in radians.

The actual involute curve, as expressed by the problem, may be drawn to scale using the AutoCad 14 program.

FIGURE 7.3 Solving involute curve coordinates with MathCad 8.

Solving for Angle φ, when the Involute Function θ is Known:

From the equation:

$$\theta = \tan\phi - \hat{\phi}$$

for $\tan\phi$, we use the identity:

$$\tan\phi = \frac{\sin\phi\,(series)}{\cos\phi\,(series)}$$

First, tan φ is set up as a sin φ/cos φ series, from inv function θ = tan (φ) - arc φ, and set equal to 0.

Next, MathCad 8 solves the following equation for angle φ, in radians, when θ = 0.0352580

$$\frac{\left[\phi - \left(\frac{\phi^3}{3!}\right) + \left(\frac{\phi^5}{5!}\right) - \left(\frac{\phi^7}{7!}\right) + \left(\frac{\phi^9}{9!}\right)\right]}{1 - \left(\frac{\phi^2}{2!}\right) + \left(\frac{\phi^4}{4!}\right) - \left(\frac{\phi^6}{6!}\right) + \left(\frac{\phi^8}{8!}\right)} - \phi - 0.0352580 = 0$$

Series equation for solving the angle φ, for a given involute function θ.

$$\begin{bmatrix} -4.9011781603967077339 - 2.3707879433169639105\cdot i \\ -4.9011781603967077339 + 2.3707879433169639105\cdot i \\ -4.1465476786869627765 \\ -.25076559865530748077 - .40920868023059962248\cdot i \\ -.25076559865530748077 + .40920868023059962248\cdot i \\ *.45922131714482631829 \\ 4.1482411156481640729 \\ 4.9016537569990014073 - 2.3723594854887197671\cdot i \\ 4.9016537569990014073 + 2.3723594854887197671\cdot i \end{bmatrix}$$

Listing of possible solutions from MathCad 8.

*0.459221317145 is the only possible value for the angle φ, in radians, for the involute function θ of 0.035258.

$0.4592213171 \cdot \left(\frac{180}{\pi}\right) = 26.31144333$ Degrees   .311·60 = 18.66   Minutes

.66·60 = 39.6   Seconds

0.4592213171 radians = φ = 26° 19' 39.6", for involute function 0.035258.

Last, we will let MathCad solve the original involute equation, by inserting the calculated angle φ, to see if this will give us the original involute function 0.035258.

φ := 0.4592213171

tan(φ) - φ = 0.03525800   Which proves the angle calculation for φ was correct.

Here we have solved the *inverse involute function*. That is, when the involute function is given or calculated, we may now find the angle for the involute function. Prior to the availability of PC programs such as MathCad, performing these operations with any degree of accuracy was next to impossible.

FIGURE 7.4 Calculating the inverse involute function in MathCad 8.

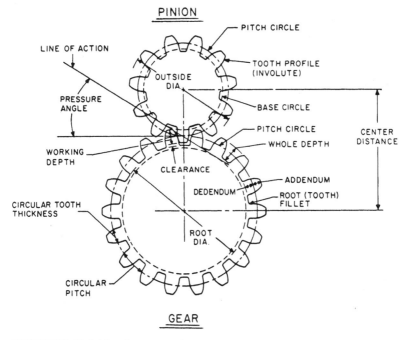

FIGURE 7.5 Definitions for spur gear terms.

For example, a U.S. customary gear of 1.6933 DP is equivalent to 1.8553 CP and to 15 module.

Proportions of standard gear teeth (U.S. customary) in relation to pitch diameter P_d are shown in Fig. 7.7.

The following figures give the formulas or equations for the different types of gear systems:

- Spur gear equations—Fig. 7.8
- Helical gear equations—Fig. 7.9
- Miter and bevel gear equations—Fig. 7.10
- Worm and worm gear equations—Fig. 7.11

Equivalent DP, CP and Module

Diametral Pitch	Circular Pitch	Module
3/4	4.1888	33.8661
0.7854	4	32.3397
0.8467	3.7106	30
1	3.1415	25.3995
1.0160	3.0922	25
1.0472	3	24.2548
1-1/4	2.5133	20.3196
1.2700	2.4737	20
1.4111	2.2264	18
1-1/2	2.0944	16.9330
1.5708	2	16.1698
1.5875	1.9790	16
1.6933	1.8553	15
1-3/4	1.7952	14.5140
1.8143	1.7316	14
1.9538	1.6079	13
2	1.5708	12.6998
2.0944	1-1/2	12.1274
2.1166	1.4842	12
2-1/4	1.3963	11.2887
2.3090	1.3606	11
2-1/2	1.2560	10.1598
2.5400	1.2369	10
2.8222	1.1132	9
3	1.0472	8.4665
3.1416	1	8.0849
3.1749	0.9895	8
3-1/2	0.8976	7.2570
3.6285	0.8658	7
4	0.7854	6.3499
4.1888	3/4	6.0637
4.2333	0.7421	6
5	0.6283	5.0799
5.0799	0.6184	5
6	0.5236	4.2333
6.2832	1/2	4.0425
6.3499	0.4947	4
8	0.3927	3.1749
8.4665	0.3711	3
10	0.3142	2.5400

FIGURE 7.6 Equivalent DP, CP, and module.

GEAR AND SPROCKET CALCULATIONS   7.9

Tooth Type	14.5° Composite	14.5° Full Depth Involute	20° Full Depth Involute	20° Stub Involute
Addendum	$1/P_d$	$1/P_d$	$1/P_d$	$0.8/P_d$
Minimum dedendum	$1.157/P_d$	$1.157/P_d$	$1.157/P_d$	$1/P_d$
Whole depth	$2.157/P_d$	$2.157/P_d$	$2.157/P_d$	$1.8/P_d$
Clearance	$0.157/P_d$	$0.157/P_d$	$0.157/P_d$	$0.2/P_d$

Note: In the composite tooth form, the middle third of the tooth profile has an involute shape, while the remainder is cycloidal.

FIGURE 7.7 Proportions of standard gear teeth.

To measure the size (diametral pitch) of standard U.S. customary gears, gear gauges are often used. A typical set of gear gauges is shown in Fig. 7.12. The measuring techniques for using gear gauges are shown in Figs. 7.13 and 7.14.

A simple planetary or epicyclic gear system is shown in Fig. 7.15a, together with the speed-ratio equations and the gear-train schematic. Extensive gear design equations and gear manufacturing methods are contained in the McGraw-Hill handbooks, *Electromechanical Design Handbook, Third Edition* (2000) and *Machining and Metalworking Handbook, Second Edition* (1999). Figure 7.15b shows an actual epicyclic gear system in a power tool. A chart of gear and sprocket mechanics equations is shown in Fig. 7.16.

To obtain	Having	Formula
Diametral pitch P	Circular pitch p	$P = \dfrac{3.1416}{p}$
	Number of teeth N and pitch diameter D	$P = \dfrac{N}{D}$
	Number of teeth N and outside diameter D_o	$t = \dfrac{1.5708}{P}$
Circular pitch p	Diametral pitch P	$p = \dfrac{3.1416}{P}$
Pitch diameter D	Number of teeth N and diametral pitch P	$D = \dfrac{N}{P}$
	Outside diameter D_o and diametral pitch P	$D = D_o - \dfrac{2}{P}$
Base diameter D_b	Pitch diameter D and pressure angle ϕ	$D_b = D \cos \phi$
Number of teeth N	Diametral pitch P and pitch diameter D	$N = P \times D$
Tooth thickness t at pitch diameter D	Diametral pitch P	$t = \dfrac{1.5708}{P}$
Addendum a	Diametral pitch P	$a = \dfrac{1}{P}$
Outside diameter D_o	Pitch diameter D and addendum a	$D_o = D + 2a$
Whole depth h_1, 20 P and finer	Diametral pitch P	$h_1 = \dfrac{2.2}{P} + 0.002$
Whole depth h_1, coarser than 20 P	Diametral pitch P	$h_1 = \dfrac{2.157}{P}$
Working depth h_k	Addendum a	$a = \dfrac{1}{P}$
Clearance c	Whole depth h_1 and addendum a	$c = h_1 - 2(a)$
Dedendum b	Whole depth h_1 and addendum a	$b = h_1 - a$
Contact ratio M_c	Outside radii, base radii, center distance C, and pressure angle ϕ	⇓
		$M_c = \dfrac{\sqrt{R_o^2 - R_b^2} + \sqrt{r_o^2 - r_b^2} - C \cos \phi}{P \cos \phi}$
Root diameter D_r	Pitch diameter D and dedendum b	$D_r = D - 2(b)$
Center distance C	Pitch diameter D or number of teeth N and pitch P	$C = \dfrac{D_1 + D_2}{2}$ or $\dfrac{N_1 + N_2}{2P}$

Note: $R_ó$ = outside radius, gear; r_o = outside radius, pinion; R_b = base circle radius, gear; r_b = base circle radius, pinion.

FIGURE 7.8 Spur gear equations.

GEAR AND SPROCKET CALCULATIONS   7.11

To obtain	Having	Formula
Transverse diametral pitch P	Number of teeth N and pitch diameter D	$P = \dfrac{N}{D}$
	Normal diametral pitch P_n and helix angle Ψ	$P = P_N \cos \psi$
Pitch diameter D	Number of teeth N and transverse diametral pitch P	$D = \dfrac{N}{P}$
Normal diametral pitch P_N	Transverse diametral pitch P and helix angle Ψ	$P_N = \dfrac{P}{\cos \psi}$
Normal circular tooth thickness τ	Normal diametral pitch P_N	$\tau = \dfrac{1.5708}{P_N}$
Transverse circular pitch p_1	Transverse diametral pitch P	$p_1 = \dfrac{\pi}{P}$
Normal circular pitch p_n	Transverse circular pitch p_1	$p_n = p_1 \cos \psi$
Lead L	Pitch diameter D and helix angle Ψ	$L = \dfrac{\pi D}{\tan \psi}$

FIGURE 7.9 Helical gear equations.

To obtain	Having	Formula Pinion	Formula Gear
Pitch diameter D, d	Number of teeth and diametral pitch P	$d = \dfrac{n}{P}$	$D = \dfrac{n}{P}$
Whole depth h_1	Diametral pitch P	$h_1 = \dfrac{2.188}{P} + 0.002$	$h_1 = \dfrac{2.188}{P} + 0.002$
Addendum a	Diametral pitch P	$a = \dfrac{1}{P}$	$a = \dfrac{1}{P}$
Dedendum b	Whole depth h_1 and addendum a	$b = h_1 - a$	$b = h_1 - a$
Clearance	Whole depth a_1 and addendum a	$c = h_1 - 2a$	$c = h_1 - 2a$
Circular tooth thickness τ	Diametral pitch P	$\tau = \dfrac{1.5708}{P}$	$\tau = \dfrac{1.5708}{P}$
Pitch angle	Number of teeth in pinion N_p and gear N	$L_p = \tan^{-1}\left(\dfrac{N_p}{N_g}\right)$	$L_g = 90 - L_p$
Outside diameter D_o, d_o	Pinion and gear pitch diameter $(D_p + D_g)$ addendum a and pitch angle $(L_p + L_g)$	$d_o = D_p + 2a(\cos L_p)$	$D_o = D_g + 2a(\cos L_g)$

FIGURE 7.10 Miter and bevel gear equations.

To obtain	Having	Formula
Circular pitch p	Diametral pitch p	$p = \dfrac{3.1416}{P}$
Diametral pitch P	Circular pitch p	$P = \dfrac{3.1416}{p}$
Lead of worm L	Number of threads in worm N_W and circular pitch p	$L = p \times N_W$
Addendum a	Diametral pitch P	$a = \dfrac{1}{P}$
Pitch diameter of worm D_W	Outside diameter d_o and addendum a	$D_W = d_o - 2(a)$
Pitch diameter of worm gear D_G	Circular pitch p and number of teeth on gear N_G	$D_G = \dfrac{N_G(p)}{3.1416}$
Center distance between worm and worm gear CD	Pitch diameter of worm D_W and worm gear D_G	$CD = \dfrac{D_W + D_G}{2}$
Whole depth of teeth h_1	Circular pitch p	$h_1 = 0.6866p$
	Diametral pitch P	$h_1 = \dfrac{2.157}{P}$
Bottom diameter of worm d_1	Whole depth h_1 and outside diameter d_W	$d_1 = d_o - 2h_1$
Throat diameter of worm gear D_1	Pitch diameter of worm gear D_G and addendum a	$D_1 = D + 2(a)$
Lead angle of worm γ	Pitch diameter of worm D_W and the lead L	$\gamma = \tan^{-1}\left(\dfrac{L}{3.1416 D_W}\right)$
Ratio	Number of teeth on gear N_G and number of threads on worm N_W	Ratio $= \dfrac{N_G}{N_W}$

FIGURE 7.11 Worm and worm gear equations.

FIGURE 7.12 A set of gear gauges.

FIGURE 7.13 Measuring miter/bevel gears.

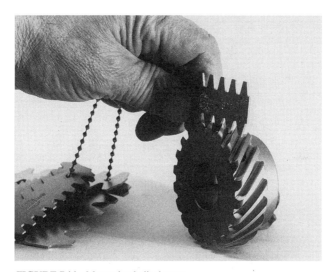

FIGURE 7.14 Measuring helical gears.

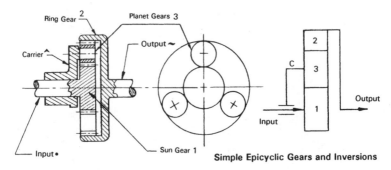

Simple Epicyclic Gears and Inversions

Input Member	Fixed Member	Output Member	Speed-ratio Equation
1	C	2	$R = -N_2/N_1$
2	C	1	$R = -N_1/N_2$
1	2	C	$R = 1 + (N_2/N_1)$
2	1	C	$R = 1 + (N_1/N_2)$
C	2	1	$R = 1/(1 + (N_2/N_1))$
C	1	2	$R = 1/(1 + (N_1/N_2))$

Note: The minus sign indicates opposite rotation from input.

(a)

FIGURE 7.15a A planetary or epicyclic gear system.

(b)

FIGURE 7.15b An actual epicyclic gear system in a power tool.

GEAR AND SPROCKET CALCULATIONS 7.15

To obtain	Having	Formula
Velocity v, ft/min	Pitch diameter D of gear or sprocket, in, and revolutions per minute (rpm)	$v = 0.2618 \times D \times \text{rpm}$
Revolutions per minute (rpm)	Velocity v, ft/min, and pitch diameter D of gear or sprocket, in	$\text{rpm} = \dfrac{v}{0.2618 \times D}$
Pitch diameter D of gear or sprocket, in	Velocity v, ft/min, and revolutions per minute (rpm)	$D = \dfrac{v}{0.2618 \times \text{rpm}}$
Torque, lb · in	Force W, lb, and radius, in	$T = W \times R$
Horsepower (hp)	Force W, lb, and velocity v, ft/min	$\text{hp} = \dfrac{W \times v}{33{,}000}$
Horsepower (hp)	Torque T, lb · in, and revolutions per minute (rpm)	$\text{hp} = \dfrac{T \times \text{rpm}}{63{,}025}$
Torque T, lb · in	Horsepower (hp) and revolutions per minute (rpm)	$T = \dfrac{63{,}025 \times \text{hp}}{\text{rpm}}$
Force W, lb	Horsepower (hp) and velocity v, ft/min	$W = \dfrac{33{,}000 \times \text{hp}}{v}$
Revolutions per minute (rpm)	Horsepower (hp) and torque T, lb · in	$\text{rpm} = \dfrac{63{,}025 \times \text{hp}}{T}$

FIGURE 7.16 Gear and sprocket mechanics equations.

7.3 SPROCKETS—GEOMETRY AND DIMENSIONING

Figure 7.17 shows the geometry of ANSI standard roller chain sprockets and derivation of the dimensions for design engineering or tool engineering use. With the following relational data and equations, dimensions may be derived for input to CNC machining centers or EDM machines for either manufacturing the different-size sprockets or producing the dies to stamp and shave the sprockets.

The equations for calculating sprockets are as follows:

P = pitch (ae)
N = number of teeth
D_r = nominal roller diameter
D_s = seating curve diameter = $1.005 D_r + 0.003$, in
$R = \frac{1}{2} D_s$
$A = 35° + (60°/N)$
$B = 18° - (56°/N)$
$ac = 0.8 D_r$
$M = 0.8 D_r \cos[(35° + (60°/N)]$
$T = 0.8 D_r \sin(35° + (60°/N))$

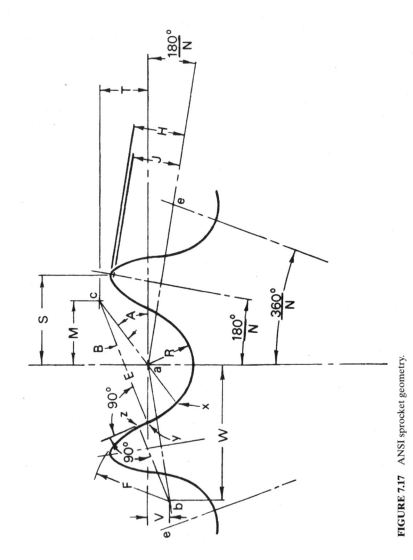

FIGURE 7.17 ANSI sprocket geometry.

GEAR AND SPROCKET CALCULATIONS

$E = 1.3025 D_r + 0.0015$, in

Chord $xy = (2.605 \, D_r + 0.003) \sin (9° - (28°/N))$, in

$yz = D_r \{1.4 \sin [17° - (64°/N)] - 0.8 \sin [18° - (56°/N)]\}$

Length of line between a and $b = 1.4 D_r$

$W = 1.4 D_r \cos (180°/N)$

$V = 1.4 D_r \sin (180°/N)$

$F = D_r \{0.8 \cos [18° - (56°/N)] + 1.4 \cos [17° - (64°/N)] - 1.3025\} - 0.0015$ in

P	D_r	R min.	D_s min.	D_s tolerance*
¼	0.130	0.0670	0.134	0.0055
⅜	0.200	0.1020	0.204	0.0055
½	0.306	0.1585	0.317	0.0060
½	0.312	0.1585	0.317	0.0060
⅝	0.400	0.2025	0.405	0.0060
¾	0.469	0.2370	0.474	0.0065
1	0.625	0.3155	0.631	0.0070
1¼	0.750	0.3785	0.757	0.0070
1½	0.875	0.4410	0.882	0.0075
1¾	1.000	0.5040	1.008	0.0080
2	1.125	0.5670	1.134	0.0085
2¼	1.406	0.7080	1.416	0.0090
2½	1.562	0.7870	1.573	0.0095
3	1.875	0.9435	1.887	0.0105

* Denotes plus tolerance only.

FIGURE 7.18 Seating curve data for ANSI roller chain (inches).

Chain number	Carbon steel, lb	Stainless steel, lb
25*	925	700
35*	2,100	1,700
40	3,700	3,000
S41	2,000	1,700
S43	1,700	—
50	6,100	4,700
60	8,500	6,750
80	14,500	12,000
100	24,000	18,750
120	34,000	27,500
140	46,000	—
160	58,000	—
180	80,000	—
200	95,000	—
240	130,000	—

* Rollerless chain.

FIGURE 7.19 Maximum loads in tension for standard ANSI chains.

ANSI STANDARD ROLLER CHAIN
Single Strand

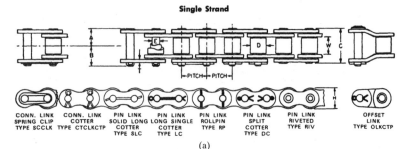

(a)

Chain number	Pitch	W	D	C	B	A	T	H	E	Weight, lb/ft
25*	¼	0.125	0.130	0.31	0.19	0.15	0.030	0.23	0.0905	0.104
35*	⅜	0.187	0.200	0.47	0.34	0.23	0.050	0.36	0.141	0.21
40	½	0.312	0.312	0.65	0.42	0.32	0.060	0.46	0.156	0.41
S41	½	0.250	0.306	0.51	0.37	0.26	0.050	0.39	0.141	0.28
S43	½	0.125	0.306	0.39	0.31	0.20	0.050	0.39	0.141	0.22
50	⅝	0.375	0.400	0.79	0.56	0.40	0.080	0.59	0.200	0.69
60	¾	0.500	0.468	0.98	0.64	0.49	0.094	0.70	0.234	0.96
80	1	0.625	0.625	0.128	0.74	0.64	0.125	0.93	0.312	1.60
100	1¼	0.750	0.750	1.54	0.91	0.77	0.156	1.16	0.375	2.56
120	1½	1.00	0.875	1.94	1.14	0.97	0.187	1.38	0.437	3.60
140	1¾	1.00	1.00	2.08	1.22	1.04	0.218	1.63	0.500	4.90
160	2	1.25	1.12	2.48	1.46	1.24	0.250	1.88	0.562	6.40
180	2¼	1.41	1.41	2.81	1.74	1.40	0.281	2.13	0.687	8.70
200	2½	1.50	1.56	3.02	1.86	1.51	0.312	2.32	0.781	10.30
240	3	1.88	1.88	3.76	2.27	1.88	0.375	2.80	0.937	16.99

* Rollerless chain.

(b)

FIGURE 7.20 ANSI standard roller chain and dimensions.

$H = \sqrt{F^2 - (1.4 D_r - 0.5 P)^2}$

$S = 0.5P \cos(180°/N) + H \sin(180°/N)$

Approximate o.d. of sprocket when J is $0.3P = P[0.6 + \cot(180°/N)]$

Outer diameter of sprocket with tooth pointed $= p \cot(180°/N) + \cos(180°N)$ $(D_s - D_r) + 2H$

Pressure angle for new chain $= xab = 35° - (120°/N)$

Minimum pressure angle $= xab - B = 17° - (64°/N)$

Average pressure angle $= 26° - (92°/N)$

The seating curve data for the preceding equations are shown in Fig. 7.18.

For maximum loads in pounds force in tension for standard ANSI chains, see Fig. 7.19. ANSI standard roller chain and dimensions are shown in Figs. 7.20a and b.

CHAPTER 8
RATCHETS AND CAM GEOMETRY

8.1 RATCHETS AND RATCHET GEARING

A *ratchet* is a form of gear in which the teeth are cut for one-way operation or to transmit intermittent motion. The ratchet wheel is used widely in machinery and many mechanisms. Ratchet-wheel teeth can be either on the perimeter of a disk or on the inner edge of a ring.

The *pawl*, which engages the ratchet teeth, is a beam member pivoted at one end, the other end being shaped to fit the ratchet-tooth flank.

Ratchet Gear Design. In the design of ratchet gearing, the teeth must be designed so that the pawl will remain in engagement under ratchet-wheel loading. In ratchet gear systems, the pawl will either push the ratchet wheel or the ratchet wheel will push on the pawl and/or the pawl will pull the ratchet wheel or the ratchet wheel will pull on the pawl. See Figs. 8.1a and b for the four variations of ratchet and pawl action. In the figure, F indicates the origin and direction of the force and R indicates the reaction direction.

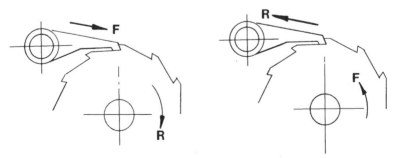

FIGURE 8.1a Variation of ratchet and pawl action (F = force; R = reaction).

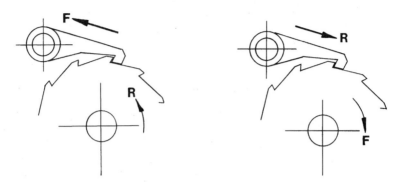

FIGURE 8.1b Variation of ratchet and pawl action (F = force; R = reaction).

Tooth geometry for case I in Fig. 8.1a is shown in Fig. 8.2. A line perpendicular to the face of the ratchet-wheel tooth must pass between the center of the ratchet wheel and the center of the pawl pivot point.

Tooth geometry for case II in Fig. 8.1b is shown in Fig. 8.3. A line perpendicular to the face of the ratchet-wheel tooth must fall outside the pivot center of the pawl and the ratchet wheel.

Spring loading the pawl is usually employed to maintain constant contact between the ratchet wheel and pawl (gravity or weight on the pawl is also sometimes used). The pawl should be pulled automatically in and kept in engagement with the ratchet wheel, independent of the spring or weight loading imposed on the pawl.

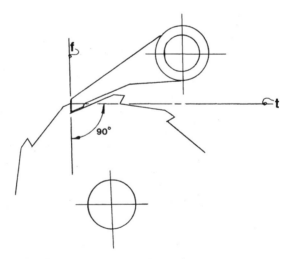

FIGURE 8.2 Tooth geometry for case I.

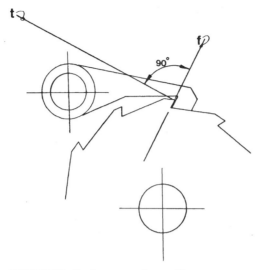

FIGURE 8.3 Tooth geometry for case II.

8.2 METHODS FOR LAYING OUT RATCHET GEAR SYSTEMS

8.2.1 External Tooth Ratchet Wheels

See Fig. 8.4.
1. Determine the pitch, tooth size, and radius R to meet the strength and mechanical requirements of the ratchet gear system (see Sec. 8.2.3, "Calculating the Pitch and Face of Ratchet-Wheel Teeth").
2. Select the position points O, O_1, and A so that they all fall on a circle C with angle OAO_1 equal to 90°.
3. Determine angle ϕ through the relationship $\tan \phi = r/c =$ a value greater than the coefficient of static friction of the ratchet wheel and pawl material—0.25 is sufficient for standard low- to medium-carbon steel. Or $r/R = 0.25$, since the sine and tangent of angle ϕ are close for angles from 0 to 30°.

NOTE. The value c is determined by the required ratchet wheel geometry; therefore, you must solve for r, so

$$r = c \tan \phi \quad \text{or} \quad r = R \tan \phi$$
$$= c(0.25) \qquad \qquad = R(0.25)$$

4. Angle ϕ is also equal to arctan (a/b), and to keep the pawl as small as practical, the center pivot point of the pawl O_1 may be moved along line t toward point A to satisfy space requirements.

8.4 CHAPTER EIGHT

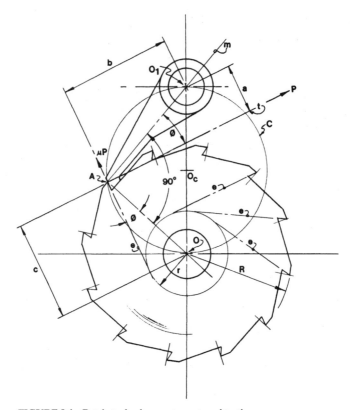

FIGURE 8.4 Ratchet wheel geometry, external teeth.

5. The pawl is then self-engaging. This follows the principle stated earlier that a line perpendicular to the tooth face must fall *between* the centers of the ratchet wheel and pawl pivot points.

8.2.2 Internal-Tooth Ratchet Wheels

See Fig. 8.5.
1. Determine the pitch, tooth size, and radii R and R_1 to meet the strength and mechanical requirements of the ratchet gear system. For simplicity, let points O and O_1 be on the same centerline.
2. Select r so that $f/g \geq 0.20$.
3. A convenient angle for β is 30°, and $\tan \beta = f/g = 0.557$, which is greater than the coefficient of static friction for steel (0.15). This makes angle $\alpha = 60°$ because $\alpha + \beta = 90°$.

NOTE. Locations of tooth faces are generated by element lines e.

RATCHETS AND CAM GEOMETRY 8.5

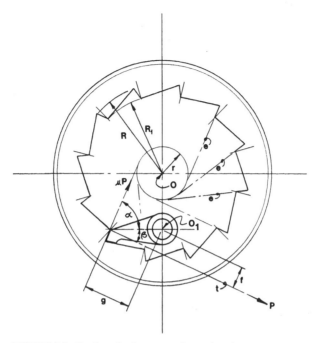

FIGURE 8.5 Ratchet wheel geometry, internal teeth.

For self-engagement of the pawl, note that a line t perpendicular to the tooth face must fall outside the pawl pivot point O_1.

8.2.3 Calculating the Pitch and Face of Ratchet-Wheel Teeth

The following equation may be used in calculating the pitch or the length of the tooth face (thickness of ratchet wheel) and is applicable to most general ratchet-wheel designs. Note that selection of the values for S_s (safe stress, psi) may be made more or less conservatively, according to the requirements of the application. Low values for S_s are selected for applications involving safety conditions. Note also that the shock stress allowable levels (psi) are 10 times less than for normal loading applications, where a safety factor is not a consideration.

The general pitch design equation and transpositions are given as

$$P = \sqrt{\frac{\alpha m}{l S_s N}} \qquad P^2 = \frac{\alpha m}{l S_s N} \qquad N = \frac{\alpha m}{l S_s P^2} \qquad l = \frac{\alpha m}{N S_s P^2}$$

where P = circular pitch measured at the outside circumference, in
  m = turning moment (torque) at ratchet-wheel shaft, lb · in
  l = length of tooth face, thickness of ratchet wheel, in
  S_s = safe stress (steel C-1018; 4000 psi shock and 25,000 psi static)

N = number of teeth in ratchet wheel
α = coefficient: 50 for 12 teeth or less, 35 for 13 to 20 teeth, and 20 for more than 20 teeth

For other materials such as brass, bronze, stainless steel, zinc castings, etc., the S_s rating may be proportioned to the values given for C-1018 steel, versus other types or grades of steels.

Laser Cutting Ratchet Wheels. A ratchet wheel cut on a wire electric discharge machine (EDM) is shown in Fig. 8.6. Note the clean, accurate cut on the teeth.

FIGURE 8.6 Ratchet wheel cut by a wire electric discharge machine (EDM).

Figure 8.7 shows the EDM that was used to cut the ratchet wheel shown in Fig. 8.6.

8.2 CAM LAYOUT AND CALCULATIONS

Cams are mechanical components which convert rotary motion into a selective or controlled translating or oscillating motion or action by way of a cam follower which bears against the working surface of the cam profile or perimeter. As the cam rotates, the cam follower rises and falls according to the motions described by the displacement curve.

Cams can be used to translate power and motion, such as the cams on the camshaft of an internal combustion engine, or for selective motions as in timing

FIGURE 8.7 The wire EDM which cut the ratchet wheel shown in Fig. 8.6.

devices or generating functions. The operating and timing cycles of many machines are controlled by the action of cams.

There are basically two classes of cams; uniform-motion cams and accelerated-motion cams.

Cam Motions. The most important cam motions and displacement curves in common use are

- *Uniform-velocity motion,* for low speeds
- *Uniform acceleration,* for moderate speeds
- *Parabolic motion used in conjunction with uniform motion or uniform acceleration,* for low to moderate speeds
- *Cycloidal,* for high speeds

The design of a typical cam is initiated with a *displacement curve* as shown in Fig. 8.8. Here, the Y dimension corresponds to the cam rise or fall, and the X dimension corresponds either to degrees, radians, or time displacement. The slope lines of the rise and fall intervals should be terminated with a parabolic curve to prevent shock loads on the follower. The total length of the displacement (X dimension) on the displacement diagram represents one complete revolution of the cam. Standard graphical layout methods may be used to develop the displacement curves and simple cam profiles. The placement of the parabolic curves at the terminations of the rise/fall intervals on uniform-motion and uniform-acceleration cams is depicted in the detail view of Fig. 8.8. The graphical construction of the parabolic curves which begin and end the rise/fall intervals may be accomplished using the principles of geometric construction shown in drafting manuals or in Chap. 3 of this book.

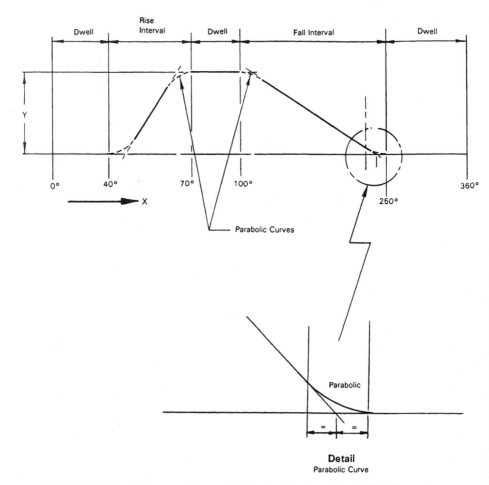

FIGURE 8.8 Cam displacement diagram (the developed cam is as shown in Fig. 8.9).

The layout of the cam shown in Fig. 8.9 is a development of the displacement diagram shown in Fig. 8.8. In this cam, we have a dwell interval followed by a uniform-motion and uniform-velocity rise, a short dwell period, a uniform fall, and then the remainder of the dwell to complete the cycle of one revolution.

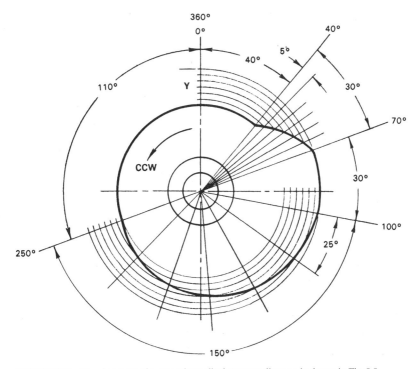

FIGURE 8.9 Development of a cam whose displacement diagram is shown in Fig. 8.8.

The layout of a cam such as shown in Fig. 8.9 is relatively simple. The rise/fall periods are developed by dividing the rise or fall into the same number of parts as the angular period of the rise and fall. The points of intersection of the rise/fall divisions with the angular divisions are then connected by a smooth curve, terminating in a small parabolic curve interval at the beginning and end of the rise/fall periods. Cams of this type have many uses in industry and are economical to manufacture because of their simple geometries.

Uniform-Motion Cam Layout. The cam shown in Fig. 8.10 is a uniform or harmonic-motion cam, often called a *heart cam* because of its shape. The layout of this type of cam is simple, as the curve is a development of the intersection of the rise intervals with the angular displacement intervals. The points of intersection are then connected by a smooth curve.

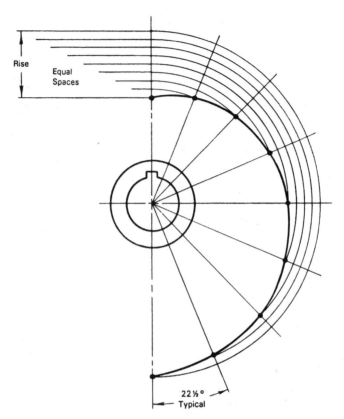

FIGURE 8.10 Uniform-motion cam layout (harmonic motion).

Accelerated-Motion Cam Layout. The cam shown in Fig. 8.11 is a uniform-acceleration cam. The layout of this type of cam is also simple. The rise interval is divided into increments of 1-3-5-5-3-1 as shown in the figure. The angular rise interval is then divided into six equal angular sections as shown. The intersection of the projected rise intervals with the radial lines of the six equal angular intervals are then connected by a smooth curve, completing the section of the cam described. The displacement diagram that is generated for the cam follower motion by the designer will determine the final configuration of the complete cam.

Cylindrical Cam Layout. A cylindrical cam is shown in Fig. 8.12 and is layed out in a similar manner described for the cams of Figs. 8.9 and 8.10. A displacement diagram is made first, followed by the cam stretchout view shown in Fig. 8.12. The points describing the curve that the follower rides in may be calculated mathematically for a precise motion of the follower. Four- and five-axis machining centers are used to cut the finished cams from a computer program generated in the engineering department and fed into the controller of the machining center.

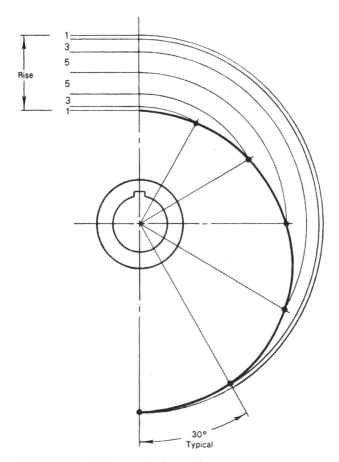

FIGURE 8.11 Uniform-acceleration cam layout.

Tracer cutting and incremental cutting are also used to manufacture cams, but are seldom used when the manufacturing facility is equipped with four- and five-axis machining centers, which do the work faster and more accurately than previously possible.

The design of cycloidal motion cams is not discussed in this handbook because of their mathematical complexity and many special requirements. Cycloidal cams are also expensive to manufacture because of the requirements of the design and programming functions required in the engineering department.

Eccentric Cams. A cam which is required to actuate a roller limit switch in a simple application or to provide a simple rise function may be made from an eccentric shape as shown in Fig. 8.13. The rise, diameter, and offset are calculated as shown in the figure. This type of cam is the most simple to design and economical to manufacture and has many practical applications. Materials used for this type of cam

8.12 CHAPTER EIGHT

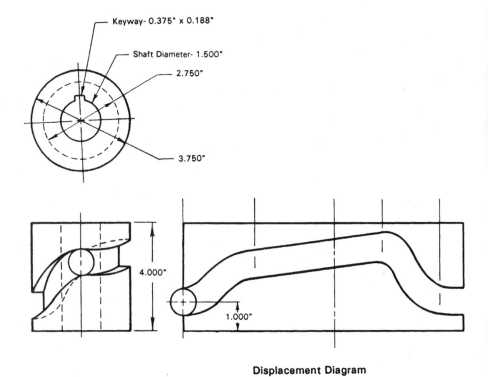

Displacement Diagram

FIGURE 8.12 Development of a cylindrical cam.

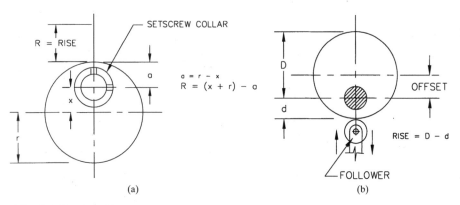

(a)   (b)

FIGURE 8.13 Eccentric cam geometry.

design can be steel, alloys, or plastics and compositions. Simple functions and light loads at low to moderate speeds are limiting factors for these types of cams.

In Fig. 8.13a and 8.13b, the simple relationships of the cam variables are as follows:

$$R = (x + r) - a \qquad a = r - x \qquad \text{rise} = D - d$$

The eccentric cam may be designed using these relationships.

The Cam Follower. The most common types of cam follower systems are the radial translating, offset translating, and swinging roller as depicted in Fig. 8.14a to 8.14c.

The cams in Figs. 8.14a and 8.14b are *open-track cams*, in which the follower must be held against the cam surface at all times, usually by a spring. A *closed-track cam* is one in which a roller follower travels in a slot or groove cut in the face of the cam. The cylindrical cam shown in Fig. 8.12 is a typical example of a closed-track cam. The closed-track cam follower system is termed positive because the follower translates in the track without recourse to a spring holding the follower against the cam surface. The positive, closed-track cam has wide use on machines in which the breakage of a spring on the follower could otherwise cause damage to the machine.

Note that in Fig. 8.14b, where the cam follower is offset from the axis of the cam, the offset must be in a direction opposite that of the cam's rotation.

On cam follower systems which use a spring to hold the cam follower against the working curve or surface of the cam, the spring must be designed properly to prevent "floating" of the spring during high-speed operation of the cam. The cyclic rate of the

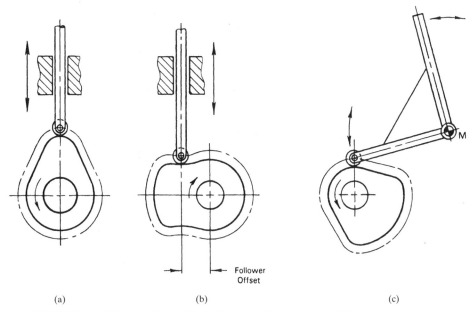

FIGURE 8.14 (a) In-line follower; (b) offset follower; (c) swinging-arm follower.

spring must be kept below the natural frequency of the spring in order to prevent floating. Chapter 10 of the handbook shows procedures for the design of high-pressure, high-cyclic-rate springs in order to prevent this phenomenon from occurring. When you know the cyclic rate of the spring used on the cam follower and its working stress and material, you can design the spring to have a natural frequency which is below the cyclic rate of operation. The placement of springs in *parallel* is often required to achieve the proper results. The valve springs on high-speed automotive engines are a good example of this practice, wherein we wish to control natural frequency and at the same time have a spring with a high spring rate to keep the engine valves tightly closed. The spring rate must also be high enough to prevent separation of the follower from the cam surface during acceleration, deceleration, and shock loads in operation. The cam follower spring is often preloaded to accomplish this.

Pressure Angle of the Cam Follower. The *pressure angle* ϕ (see Fig. 8.15) is generally made 30° or less for a reciprocating cam follower and 45° or less for an oscillating cam follower. These typical pressure angles also depend on the cam mechanism design and may be more or less than indicated above.

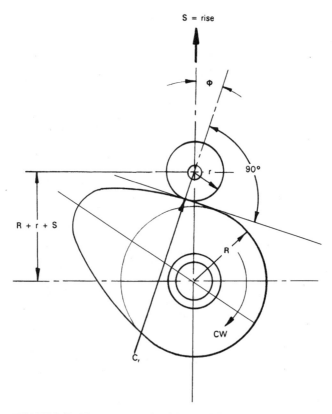

FIGURE 8.15 The pressure angle of the cam follower.

The pressure angle ϕ is the angle between a common *normal* to both the roller and the cam profile and the direction of the follower motion, with one leg of the angle passing through the axis of the follower roller axis. This pressure angle is easily found using graphical layout methods.

To avoid undercutting cams with a roller follower, the radius r of the roller must be less than C_r, which is the minimum radius of curvature along the cam profile.

Pressure Angle Calculations. The pressure angle is an important factor in the design of cams. Variations in the pressure angle affect the transverse forces acting on the follower.

The simple equations which define the maximum pressure angle α and the cam angle θ at α are as follows (see Fig. 8.16a):

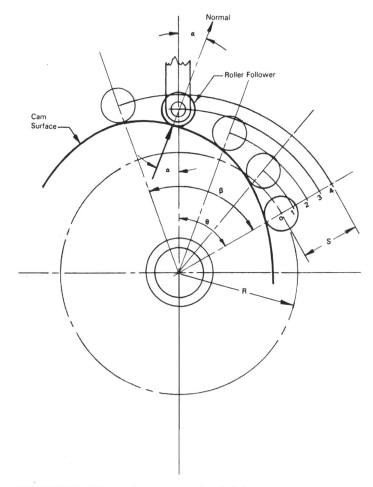

FIGURE 8.16a Diagram for pressure angle calculations.

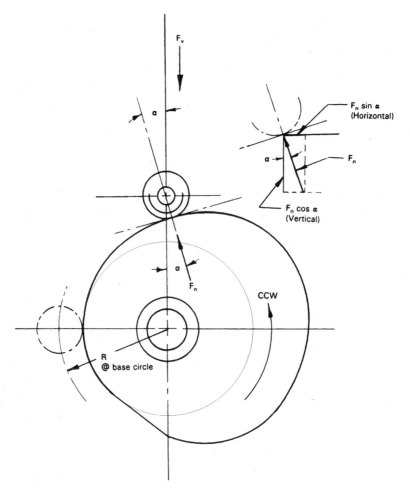

FIGURE 8.16b Normal load diagram and vectors, cam, and follower.

For simple harmonic motion:

$$\alpha = \arctan \frac{\pi}{2\beta}\left(\frac{S/R}{\sqrt{1+(S/R)}}\right) \qquad \theta = \frac{\beta}{\pi}\arccos\left(\frac{S/R}{2+(S/R)}\right)$$

For constant-velocity motion:

$$\alpha = \arctan \frac{1}{\beta}\left(\frac{S}{R}\right) \qquad \theta = 0$$

For constant-acceleration motion:

$$\alpha = \arctan \frac{2}{\beta}\left(\frac{S/R}{1+(S/R)}\right) \qquad \theta = \beta$$

For cycloidal motion:

$$\alpha = \arctan \frac{1}{2\beta}\left(\frac{S}{R}\right) \quad \theta = 0$$

where α = maximum pressure angle of the cam, degrees
  S = total lift for a given cam motion during cam rotation, in
  R = initial base radius of cam; center of cam to center of roller, in
  β = cam rotation angle during which the total lift S occurs for a given cam motion, rad
  θ = cam angle at pressure angle α

Contact Stresses Between Follower and Cam. To calculate the approximate stress S_s developed between the roller and the cam surface, we can use the simple equation

$$S_s = C\sqrt{\frac{f_n}{w}\left(\frac{1}{r_f} + \frac{1}{R_c}\right)}$$

where C = constant (2300 for steel to steel; 1900 for steel roller and cast-iron cam)
  S_s = calculated compressive stress, psi
  f_n = normal load between follower and cam surface, lbf
  w = width of cam and roller common contact surface, in
  R_c = minimum radius of curvature of cam profile, in
  r_f = radius of roller follower, in

The highest stress is developed at the minimum radius of curvature of the cam profile. The calculated stress S_s should be less than the maximum allowable stress of the weaker material of the cam or roller follower. The roller follower would normally be the harder material.

Cam or follower failure is usually due to fatigue when the surface endurance limit (permissible compressive stress) is exceeded.

Some typical maximum allowable compressive stresses for various materials used for cams, when the roller follower is hardened steel (Rockwell C45 to C55) include

Gray iron—cast (200 Bhn) ASTM A48-48	55,000 psi
SAE 1020 steel (150 Bhn)	80,000 psi
SAE 4150 steel HT (300 Bhn)	180,000 psi
SAE 4340 steel HT (R_c 50)	220,000 psi

NOTE. Bhn designates Brinnel hardness number; R_c is Rockwell C scale.

Cam Torque. As the follower bears against the cam, resisting torque develops during rise S, and assisting torque develops during fall or return. The maximum torque developed during cam rise operation determines the cam drive requirements.

The instantaneous torque values T_i may be calculated using the equation

$$T_i = \frac{9.55 y F_n \cos \alpha}{N}$$

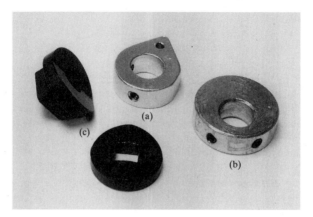

FIGURE 8.17 Typical simple cams: (*a*) quick-rise cam; (*b*) eccentric cam; (*c*) set of special rotary profile cams.

where T_i = instantaneous torque, lb · in
v = velocity of follower, in/sec
F_n = normal load, lb
α = maximum pressure angle, degrees
N = cam speed, rpm

The normal load F_n may be found graphically or calculated from the vector diagram shown in Fig. 8.16*b*. Here, the horizontal or lateral pressure on the follower = F_n sin α and the vertical component or axial load on the follower = F_n cos α.

When we know the vertical load (axial load) on the follower, we solve for F_n (the normal load) on the follower from

$$F_n \cos \alpha = F_v$$

given α = pressure angle, degrees
F_v = axial load on follower (from preceding equation), lbf
F_n = normal load at the cam profile and follower, lbf

EXAMPLE. Spring load on the follower is 80 lb and the pressure angle α is 17.5°. Then

$$F_v = F_n \cos \alpha \qquad F_n = \frac{F_v}{\cos \alpha} = \frac{80}{\cos 17.5} = \frac{80}{0.954} = 84 \text{ lb}$$

Knowing the normal force F_n, we can calculate the pressure (stress) in pounds per square inch between the cam profile and roller on the follower (see Fig. 8.16*b*).

Figure 8.17 shows typical simple cams.

CHAPTER 9
BOLTS, SCREWS, AND THREAD CALCULATIONS

9.1 PULLOUT CALCULATIONS AND BOLT CLAMP LOADS

Screw thread systems are shown with their basic geometries and dimensions in Sec. 5.2.

Engagement of Threads. The length of engagement of a stud end or bolt end E can be stated in terms of the major diameter D of the thread. In general,

- For a steel stud in cast iron or steel, $E = 1.50D$.
- For a steel stud in hardened steel or high-strength bronze, $E = D$.
- For a steel stud in aluminum or magnesium alloys subjected to shock loads, $E = 2.00D + 0.062$.
- For a steel stud as described, subjected to normal loads, $E = 1.50D + 0.062$.

Load to Break a Threaded Section. For screws or bolts,

$$P_b = SA_{ts}$$

where P_b = load to break the screw or bolt, lbf
S = ultimate tensile strength of screw or bolt material, lb/in^2
A_{ts} = tensile stress area of screw or bolt thread, in^2

NOTE. UNJ round-root threads will develop higher loads and have higher endurance limits.

Tensile Stress Area Calculation. The tensile stress area A_{ts} of screws and bolts is derived from

$$A_{ts} = \frac{\pi}{4}\left(D - \frac{0.9743}{n}\right)^2 \quad \text{(for inch-series threads)}$$

where A_{ts} = tensile stress area, in^2
D = basic major diameter of thread, in
n = number of threads per inch

NOTE. You may select the stress areas for unified bolts or screws by using Figs. 9.5 and 9.6 in Sec. 9.3, while the metric stress areas may be derived by converting millimeters to inches for each metric fastener and using the preceding equation.

Thread Engagement to Prevent Stripping. The calculation approach depends on materials selected.

1. *Same materials* chosen for both external threaded part and internal threaded part:

$$E_L = \frac{2A_{ts}}{\pi D_m \left\{ \frac{1}{2} + [n(p_d - D_m)/\sqrt{3}] \right\}}$$

where E_L = length of engagement of the thread, in
D_m = maximum minor diameter of internal thread, in
n = number of threads per in
A_{ts} = tensile stress area of screw thread as given in previous equation
p_d = minimum pitch diameter of external thread, in

2. *Different materials;* i.e., internal threaded part of lower strength than external threaded part:
   a. Determine relative strength of external thread and internal thread from

$$R = \frac{A_{se}(S_e)}{A_{si}(S_i)}$$

where R = relative strength factor
A_{se} = shear area of external thread, in^2
A_{si} = shear area of internal thread, in^2
S_e = tensile strength of external thread material, psi
S_i = tensile strength of internal thread material, psi

   b. If R is ≤ 1, the length of engagement as determined by the equation in item 1 (preceding) is adequate to prevent stripping of the internal thread. If R is > 1, the length of engagement G to prevent internal thread strip is

$$G = E_L R$$

In the immediately preceding equation, A_{se} and A_{si} are the shear areas and are calculated as follows:

$$A_{se} = \pi n E_L D_m \left[\frac{1}{2n} + \frac{(p_d - D_m)}{\sqrt{3}} \right]$$

$$A_{si} = \pi n E_L D_M \left[\frac{1}{2n} + \frac{(D_M - D_p)}{\sqrt{3}} \right]$$

where D_p = maximum pitch diameter of internal thread, in
D_M = minimum major diameter of external thread, in (Other symbols have been defined previously.)

Thread Engagement to Prevent Stripping and Bolt Clamp Loads

Problem. What is the minimum length of thread engagement required to prevent stripping threads for the following conditions:

1. Bolt size = 0.375-16 UNC-2A.
2. Torque on bolt = 32 lb · ft.
3. Internal threads will be in aluminum alloy, type 2024-T4.

Solution. From condition 2, the clamp load L developed by the bolt is calculated from:

$$T = KLD \qquad L = \frac{T}{KD}$$

Given: $K = 0.15$, $D = 0.375$ in, $T = 32 \times 12 = 384$ lb · in

$$L = \frac{384}{0.15 \times 0.375}$$

$$L = \frac{384}{0.5625} = 6827 \text{ lbf}$$

We have two different materials involved: (1) a steel bolt and (2) internal threads in aluminum alloy. So, we need to determine the relative strength factor R of the materials from the following equation (see previous symbols):

$$R = \frac{A_{se}(S_e)}{A_{si}(S_i)}$$

Next, we need to find the effective engagement length E_L from the following equation:

$$E_L = \frac{2A_{ts}}{\pi D_m \left\{ \frac{1}{2} + [n(P_d - D_m)/\sqrt{3}] \right\}}$$

where A_{ts} for 0.375-16 bolt = 0.0775 in²
$D_m = 0.321$ in
$P_d = 0.3287$ in
$n = 16$

NOTE. For A_{ts}, see thread data table or calculate tensile stress area from previous equation.

$$E_L = \frac{2 \times 0.0775}{3.1416 \times 0.321\{0.500 + [16(0.3287 - 0.321)/1.732]\}}$$

$$E_L = \frac{0.155}{1.008451(0.57113)}$$

$$E_L = \frac{0.1550}{0.5760} = 0.269 \text{ in}$$

NOTE. If E_L seems low in value, consider the facts that a 0.375-16 UNC steel hex nut is 0.337 in thick, that the jamb nut in this size is only 0.227 in thick, and that these nuts are designed so that the bolt will break *before* the threads will strip.

Next, calculate A_{se} and A_{si} from the following:

$$A_{se} = \pi n E_L D_m \left[\frac{1}{2n} + \frac{(P_d - D_m)}{\sqrt{3}} \right]$$

$$A_{se} = 4.4304(0.03125 + 0.00445)$$

$$A_{se} = 0.158 \text{ in}^2$$

where $D_m = 0.321$
$P_d = 0.3287$
$E_L = 0.269$

and

$$A_{si} = \pi n E_L D_m \left[\frac{1}{2n} + \frac{(D_M - D_p)}{\sqrt{3}} \right]$$

$$A_{si} = 4.8610(0.03125 + 0.01068)$$

$$A_{si} = 0.204 \text{ in}^2$$

where $D_M = 0.3595$
$D_p = 0.3401$

Next, use materials tables to find ultimate or tensile strength of a grade 5, 0.375-16 UNC-2A bolt, and the ultimate or tensile strength of 2024-T4 aluminum alloy:

$$S_e = 120,000 \text{ psi for grade 5 bolt}$$

$$S_i = 64,000 \text{ psi for 2024-T4 aluminum alloy}$$

and

$$R = \frac{A_{se}(S_e)}{A_{si}(S_i)}$$

$$R = \frac{0.158(120,000)}{0.204(64,000)}$$

$$R = \frac{18,960}{13,056} = 1.452$$

BOLTS, SCREWS, AND THREAD CALCULATIONS 9.5

Per the text, if R is greater than 1 (≥ 1), the adjusted length of engagement G is:

$$G = E_L R$$
$$G = 0.269 \times 1.452$$
$$G = 0.391 \text{ in} \quad \text{(adjusted length of engagement)}$$

Therefore, the *minimum* length of thread engagement for a grade 5 steel bolt tightened into a tapped hole in 2024-T4 aluminum alloy is 0.391 in. In practice, an additional 0.06 in should be added to 0.391 in, to allow for imperfect threads on the end of the bolt, thereby arriving at the final length of 0.451 in. This would then be the minimum amount of thread engagement allowed into the aluminum alloy part that would satisfy the conditions of the problem.

9.2 MEASURING AND CALCULATING PITCH DIAMETERS OF THREADS

Calculating the Pitch Diameter of Unified (UN) and Metric (M) Threads. It is often necessary to find the pitch diameter of the various unified (UN) and metric (M) thread sizes. This is necessary for threads that are not listed in the tables of thread sizes in Sec. 9.3 and when the thread is larger than that normally listed in handbooks. These include threads on large bolts and threads on jack screws and lead screws used on various machinery or machine tools. In order to calculate the pitch diameters, refer to Fig. 9.1.

$$H = 0.5\sqrt{3} \cdot p = 0.866025 p$$

where p = pitch of the thread. In the UN system, this is equal to the reciprocal of the number of threads per inch (i.e., for a ⅜-16 thread the pitch would be ¹⁄₁₆ = 0.0625 in). For the M system, the pitch is given in millimeters on the thread listing (i.e., on an M12 × 1.5 metric thread, the pitch would be 1.5 mm or 1.5 × 0.03937 in = 0.059055 in).

d = basic diameter of the external thread (i.e., ⅜-16 would be 0.375 in; #8-32 would be 0.164 in, etc.).

EXAMPLE. Find the pitch diameter of a 0.375-16 UNC-3A thread.
Using Fig. 9.1,

d = basic outside diameter of the thread = 0.375 in

$H = 0.866025 \times p = 0.866025 \times 0.0625 = 0.054127$ in (for this case only)

We would next perform the following:

$$\text{Pitch dia.} = \left(\frac{d}{2} - \frac{5H}{8} + \frac{H}{4}\right) \times 2$$

$$= \left[\frac{0.375}{2} - \left(5 \times \frac{0.054127}{8}\right) + \left(\frac{0.054127}{4}\right)\right] \times 2$$

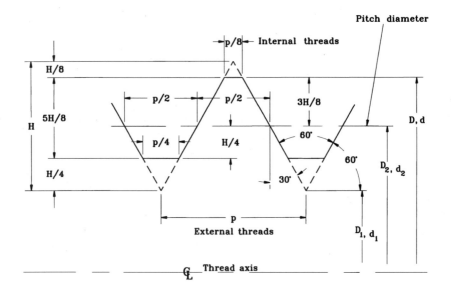

D, (d) = basic major diameter of internal (external) thread
D_1, (d_1) = basic minor diameter of internal (external) thread
D_2, (d_2) = basic pitch diameter of internal (external) thread
p = pitch
H = 0.5√3 p

FIGURE 9.1 Basic thread profile for unified (UN) and metric (M) threads (ISO 68).

$$= (0.1875 - 0.033829 + 0.013532) \times 2$$

$$= 0.3344 \text{ in pitch dia. for a 3/8-16 UNC-3A thread}$$

If you check the basic pitch diameter for this thread in a table of pitch diameters, you will find that this is the correct answer when the thread is class 3A and the pitch diameter is maximum. Thus, you may calculate any pitch diameter for the different classes of fits on any UN- or M-profile thread, since the thread geometry is shown in Fig. 9.1. Pitch diameters for other classes or types of thread systems may be calculated when you know the basic thread geometry, as in this case for the UN and M thread systems. (See Chap. 5.)

The various thread systems used worldwide include ISO-M and UN, UNJ (controlled root radii), Whitworth (BSW), American Buttress (7° face), NPT (American National Pipe Thread), BSPT (British Standard Pipe Thread), Acme (29°), Acme (stub 29°), API (taper 1:6), TR DIN 103, and RD DIN 405 (round). The geometry of all these systems is shown in Sec. 5.2.

Three-Wire Method for Measuring the Pitch Diameter of V and Acme Threads. See Fig. 9.2.

Problem. Determine the measurement *M* over three wires, and confirm the accuracy of the pitch diameter for given sizes and angles of V threads and 29° Acme threads.

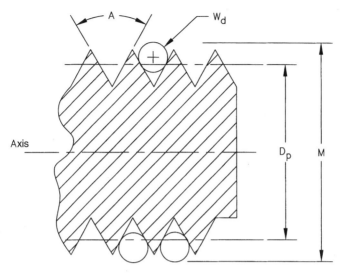

FIGURE 9.2 Three-wire method for measuring pitch diameter.

Solution. There are three useful equations for measuring over three wires to determine the pitch diameter of the different thread systems, in all classes of fits. Following are the application data for using the three equations.

1. The Buckingham simplified equation includes the effect of the screw thread lead angle, for good results on V threads with small lead angles.

$$M = D_p + W_d (1 + \sin A_n) \qquad W_d = \frac{T \cos B}{\cos A_n} = \text{required wire size} \quad \text{(Eq. 9.1)}$$

2. For *very good* accuracy, the following equation is used by the National Institute of Standards Technology (NIST), taking the lead angle into consideration:

$$M = D_p - T \cot A + W_d (1 + \csc A + 0.5 \tan^2 B \cos A \cot A) \quad \text{(Eq. 9.2)}$$

Transposed for D_p:

$$D_p = T \cot A - W_d (1 + \csc A + 0.5 \tan^2 B \cos A \cot A) + M$$

3. For *very high* accuracy for the measured value of M, use the Buckingham exact involute helicoid equation applied to screw threads:

$$M = \frac{2R_b}{\cos G} + W_d \quad \text{(Eq. 9.3)}$$

Auxiliary equations required for solving Eq. 9.3 include Eqs. 9.3a through 9.3f:

$$\tan F = \frac{\tan A}{\tan B} = \frac{\tan A_n}{\sin B} \quad \text{(Eq. 9.3a)}$$

9.8 CHAPTER NINE

$$R_b = \frac{D_p}{2} \cos F \qquad \text{(Eq. 9.3}b\text{)}$$

$$T_a = \frac{T}{\tan B} \qquad \text{(Eq. 9.3}c\text{)}$$

$$\tan H_b = \cos F \tan H \qquad \text{(Eq. 9.3}d\text{)}$$

$$\text{inv } G = \frac{T_a}{D_p} + \text{inv } F + \frac{W_d}{2R_b \cos H_b} - \frac{\pi}{S} \qquad \text{(Eq. 9.3}e\text{)}$$

$$W_d = \frac{T \cos B}{\cos A_n} \qquad \text{(Eq. 9.3}f\text{)}$$

NOTE. $H = 90° - B$

Symbols for Eqs. 9.1, 9.2, 9.3, and 9.3a to 9.3f

B = lead angle at pitch diameter = helix angle; $\tan B = L/\pi D_p$
D_p = pitch diameter for which M is required, or pitch diameter according to the M measurement
A = ½ included thread angle in the axial plane
A_n = ½ included thread angle in the plane perpendicular to the sides of the thread; $\tan A_n = \tan A \cos B$
L = lead of the thread = pitch × number of threads or leads (i.e., pitch × 2 for two leads)
M = measurement over three wires per Fig. 9.2
p = pitch = 1/number of threads per inch (U.S. customary) or per mm (metric)
$T = 0.5p$ = width of thread in the axial plane at the pitch diameter
T_a = arc thickness on pitch circle on a plane perpendicular to the axis (calculate from Eq. 9.3c)
W_d = wire diameter for measuring M (see Eqs. 9.3f and 9.4)
H = helix angle at the pitch diameter from axis = $90° - B$ or $\tan H = \cot B$
H_b = helix angle at R_b measured from axis (calculate from Eq. 9.3d)
F = angle required for Eq. 9.3 group (calculate from Eq. 9.3a)
G = angle required for Eq. 9.3 group
R_b = radius required for Eq. 9.3 group (calculate from Eq. 9.3b)
S = number of *starts* or threads on a multiple-thread screw (used in Eq. 9.3e)

Equations for Determining Wire Sizes. For precise results:

$$W = \frac{T \cos B}{\cos A_n} \qquad \text{(Eq. 9.3}f\text{)}$$

For good results:

$$W = \frac{T}{\cos A} \qquad \text{(Eq. 9.4)}$$

Use Eq. 9.2 for *best size* commercial wire which makes contact at or very near the pitch diameter. Use Eq. 9.1 for relatively large lead angles, using special wire sizes as calculated from the wire size equations. Use Eq. 9.3 for precise accuracy, using the wire sizes calculated from Eq. 9.3f.

Problem. What should be the *nominal M* measurement for a class 2A, 0.500-13 UNC thread?
Solution. See Fig. 9.2.

Step 1. Select the equation (9.1, 9.2, or 9.3) for the accuracy required.

Step 2. Measure M using commercial wire size or wire size calculated from Eq. 9.3f or 9.4.

Step 3. Calculate M using the selected equation for the required pitch diameter accuracy. Then determine the tolerance of the calculated M to the measured M for the class of thread being checked, using a table of screw thread standard dimensional limits for pitch diameters.

Problem. How do you find the *actual* machined pitch diameter of a thread specified as 0.3125-18 UNC, class 1, for a particular measurement of the M dimension shown in Fig. 9.2?
Solution. See Fig. 9.2.

Step 1. Select the correct wire size and measure the M dimension of the thread being checked.

Step 2. Use Eq. 9.2 in its transposed form and calculate the actual pitch diameter D_p per the measurement M, taken across three wires as shown in Fig. 9.2.

Step 3. Check the thread table value of the pitch diameter limits, to see if the calculated pitch diameter of the thread size being checked is within acceptable tolerances or specifications.

Measuring M, *Checking Pitch Diameter, and Calculating Wire Size (New Method).* Calculate the measurement M over three wires, to confirm the accuracy of the pitch diameter for a given size of V thread (see Fig. 9.2).
Using the Buckingham simplified equation:

$$M = D_p + W_d (1 - \sin A_n)$$

where $W_d = T \cos B / \cos A_n$
$\tan B = L / \pi D_p$
$\tan A_n = \tan A \cos B$
L = pitch × no. of leads
D_p = mean or average pitch diameter

(See symbols given for previous equations.)
Given: Thread size = 0.500-13 UNC-2A; mean pitch diameter = 0.4460 in (from table of threads); pitch = 1/13 = 0.076923 in

$$\tan B = \frac{0.076923}{3.1416 \times 0.4460}$$

$$\tan B = 0.0549$$

$$\arctan 0.0549 = 3.1424° = \text{angle } B$$

$$\tan A_n = \tan A \cos B$$

$$\tan A_n = \tan 30° \times \cos 3.1424°$$

$$\tan A_n = 0.57735 \times 0.99850$$

$$\tan A_n = 0.5765$$

$$\arctan 0.5765 = 29.9634° = \text{angle } A_n$$

Then, calculate the wire diameter from:

$$W_d = \frac{T \cos B}{\cos A_n}$$

$$W_d = \frac{0.5(1/13) \cos 3.1424°}{\cos 29.9634°}$$

$$W_d = \frac{0.03840}{0.86634}$$

$$W_d = 0.04432 \text{ in}$$

Next, calculate M from:

$$M = D_p + W_d(1 - \sin A_n)$$

$$M = 0.4460 + 0.04432(1 + \sin 29.9634°)$$

$$M = 0.4460 + 0.06646$$

$$M = 0.5125 \text{ in}$$

The wire diameter W_d can also be determined by using a scale AutoCad drawing of the V thread, as shown in Fig. 9.3.

The AutoCad drawing was made using a scale of 10:1, and then AutoCad measured the diameter of the wire. It measured the wire diameter as 0.0447 in, while the diameter was calculated previously as 0.04432 in. That is a difference of only 0.0004 in, which is sufficient for moderate accuracy, and indicates a low thread lead angle, as found on single-lead V threads. Acme 29° standard and stub threads may also be measured in this manner, when the thread geometry is known. See Sec. 5.2 for

BOLTS, SCREWS, AND THREAD CALCULATIONS   9.11

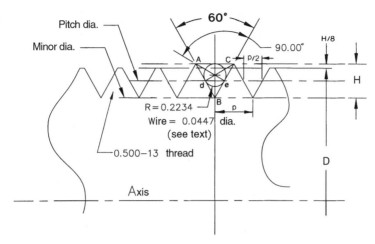

FIGURE 9.3   AutoCad scale drawing of V thread.

the geometry of international thread systems, including buttress, Acme, Whitworth 55°, etc.

A *new method* for calculating the wire diameter needed to check the accuracy of 60° V threads is as follows. As shown in Fig. 9.4, the triangle ABC is equilateral, all sides being equal. This shows that the slope lengths of the thread teeth are equal to the pitch p of the given thread. Since the circle within the triangle ABC is tangent to the sides of the triangle, we may calculate the diameter of the circle (wire diameter) as follows (see Fig. 2.10):

$$r = \frac{\sqrt{s(s-a)(a-b)(s-c)}}{s}$$

where $s = \dfrac{a+b+c}{2}$

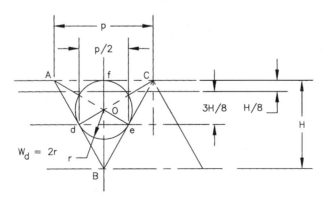

FIGURE 9.4   New method for calculating the wire diameter.

In the triangle ABC of Fig. 9.4, $a = b = c =$ pitch p, and $s = 3(p)/2$. Therefore, the equation may be rewritten as:

$$r = \frac{\sqrt{s(s-p)^3}}{s}$$

where $p =$ pitch

$W_d = 2r$

which is the new working equation for finding the wire diameter W_d of 60° V threads.

If we wish to find the wire diameter W_d in order to calculate the M dimension and check the pitch diameter accuracy of a 0.750-10 UNC-2A thread, we can use the preceding simplified equation for calculating the appropriate wire size, as follows:

Given: $p =$ pitch $= 1/10 = 0.10$ in; $s = 3 \times 0.10/2 = 0.150$
Then:

$$r = \frac{\sqrt{s(s-p)^3}}{s}$$

$$r = \frac{\sqrt{0.150(0.150 - 0.10)^3}}{0.150}$$

$$r = \frac{\sqrt{0.00001875}}{0.150} = \frac{0.00433}{0.150}$$

$$r = 0.02887$$

and $\quad W_d = 2 \times 0.02887 = 0.0577$ in

The wire diameter for calculating the M dimension would then be 0.0577 in.

You may check this diameter of 0.0577 in against the calculated diameter using the previous equation

$$W_d = \frac{T \cos B}{\cos A_n}$$

which requires one to first calculate the angles B and A_n and the width T for the 0.750-10 UNC-2A thread. The difference between the wire diameters calculated using both methods will be negligibly small. So, to save time, the new equation for calculating r and W_d may be used in conjunction with the Buckingham simplified equation for M.

The calculated wire diameter W_d for checking the pitch diameter of the 0.750-10 UNC-2A thread using the preceding equation is 0.0576 in. So, the difference in calculated wire size between the two methods shown is $0.0577 - 0.0576 = 0.0001$ in. As can be seen, the difference is indeed negligible for all but the most precision work involving 60° V threads.

BOLTS, SCREWS, AND THREAD CALCULATIONS 9.13

9.3 THREAD DATA (UN AND METRIC) AND TORQUE REQUIREMENTS (GRADES 2, 5, AND 8 U.S. STANDARD 60° V)

Figure 9.5 shows data for UNC (coarse) threads.
Figure 9.6 shows data for UNF (fine) threads.
Figure 9.7 shows data for metric M-profile threads.
Table 9.1 shows recommended tightening torques for U.S. UN SAE grade 2, 5, and 8 bolts.

Thread	Tap drill	Decimal, in	Stress area, in^2	Basic pitch diameter,
#1–64	#53	0.0595	0.0026	0.0629
#2–56	#50	0.0700	0.0037	0.0744
#3–48	#47	0.0785	0.0048	0.0855
#4–40	#43	0.0890	0.0060	0.0958
#5–40	#38	0.1015	0.0080	0.1088
#6–32	#36	0.1065	0.0090	0.1177
#8–32	#29	0.1360	0.0140	0.1437
#10–24	#25	0.1495	0.0175	0.1629
¼–20	#7	0.2010	0.0318	0.2175
⁵⁄₁₆–18	F	0.2570	0.0524	0.2764
⅜–16	⁵⁄₁₆	0.3125	0.0775	0.3344
⁷⁄₁₆–14	T	0.3580	0.1063	0.3911
½–13	²⁷⁄₆₄	0.4219	0.1419	0.4500
⁹⁄₁₆–12	³¹⁄₆₄	0.4844	0.1820	0.5084
⅝–11	¹⁷⁄₃₂	0.5312	0.2260	0.5660
¾–10	⁴¹⁄₆₄	0.6406	0.3340	0.6850
⅞–9	⁴⁹⁄₆₄	0.7656	0.4620	0.8028
1–8	⅞	0.8750	0.6060	0.9188

FIGURE 9.5 Screw thread data, Unified National Coarse (UNC).

Thread	Tap drill	Decimal, in	Stress area, in^2	Basic pitch diameter, in
#0–80	3/64	0.0469	0.0018	0.0519
#1–72	#53	0.0595	0.0027	0.0640
#2–64	#50	0.0700	0.0039	0.0759
#3–56	#45	0.0820	0.0052	0.0874
#4–48	#42	0.0935	0.0066	0.0985
#5–44	#37	0.1040	0.0083	0.1102
#6–40	#33	0.1130	0.0102	0.1218
#8–36	#29	0.1360	0.0147	0.1460
#10–32	#21	0.1590	0.0200	0.1697
1/4–28	#3	0.2130	0.0364	0.22268
5/16–24	I	0.2720	0.0580	0.2854
3/8–24	Q	0.3320	0.0878	0.3479
7/16–20	25/64	0.3906	0.1187	0.4050
1/2–20	29/64	0.4531	0.1599	0.4675
9/16–18	33/64	0.5156	0.2030	0.5264
5/8–18	9/16	0.5625	0.2560	0.5889
3/4–16	11/16	0.6875	0.3730	0.7094
7/8–14	13/16	0.8125	0.5090	0.8286
1–12	29/32	0.9063	0.6630	0.9459

FIGURE 9.6 Screw thread data, Unified National Fine (UNF).

BOLTS, SCREWS, AND THREAD CALCULATIONS 9.15

Thread designation dia × pitch, mm	Tap drill, mm	Pitch dia. 6H, internal, mm	Pitch dia. 6G, external, mm
M1.6 × 0.35	1.25	1.373	1.291
M2 × 0.4	1.60	1.740	1.654
M2.5 × 0.45	2.05	2.208	2.117
M3 × 0.5	2.50	2.675	2.580
M3.5 × 0.6	2.90	3.110	3.004
M4 × 0.7	3.30	3.545	3.433
M5 × 0.8	4.20	4.480	4.361
M6 × 1	5.00	5.350	5.212
M8 × 1.25	6.70	7.188	7.042
M8 × 1	7.00	7.350	7.212
M10 × 1.5	8.50	9.026	8.862
M10 × 1.25	8.70	9.188	9.042
M10 × 0.75	—	9.513	9.391
M12 × 1.75	10.20	10.863	10.679
M12 × 1.5	—	11.026	10.854
M12 × 1.25	10.80	11.188	11.028
M12 × 1	—	11.350	11.206
M14 × 2	12.00	12.701	12.503
M14 × 1.5	12.50	13.026	12.854
M15 × 1	—	14.350	14.206
M16 × 2	14.00	14.701	14.503
M16 × 1.5	14.50	15.026	14.854
M17 × 1	—	16.350	16.206
M18 × 1.5	16.50	17.026	16.854
M20 × 2.5	17.50	18.376	18.164
M20 × 1.5	18.50	19.026	18.854
M20 × 1	—	19.350	19.206
M22 × 2.5	19.50	20.376	20.164
M22 × 1.5	20.50	21.026	20.854
M24 × 3	21.00	22.051	21.803
M24 × 2	22.00	22.701	22.493
M25 × 1.5	—	24.026	23.854

FIGURE 9.7 Metric thread data, M profile, internal and external.

Bolt size	SAE grade 2 Tightening torque range, lb·ft	SAE grade 2 Clamp load range, lb	SAE grade 5 Tightening torque range, lb·ft	SAE grade 5 Clamp load range, lb	SAE grade 8 Tightening torque range, lb·ft	SAE grade 8 Clamp load range, lb
1/4–20	5.7–4.3	1,813–1,360	9.1–6.9	2,926–2,195	12.9–9.7	4,134–3,101
1/4–28	6.5–4.9	2,075–1,556	10.5–7.9	3,349–2,512	14.8–11.1	4,732–3,549
5/16–18	11.7–8.8	2,987–2,240	18.8–14.1	4,821–3,616	26.2–20.0	6,812–5,109
5/16–24	12.9–9.7	3,306–2,480	20.8–15.6	5,336–4,002	29.5–22.1	7,540–5,655
3/8–16	20.7–15.5	4,418–3,314	33.4–25.1	7,130–5,348	47.2–35.4	10,075–7,556
3/8–24	23.5–17.6	5,005–3,754	37.9–28.4	8,078–6,059	53.5–40.1	11,414–8,561
7/16–14	33.1–24.9	6,059–4,544	53.5–40.1	9,780–7,335	75.6–56.7	13,819–10,364
7/16–20	37.0–27.8	6,766–5,075	59.7–44.8	10,920–8,190	84.4–63.3	15,431–11,573
1/2–13	50.6–37.9	8,088–6,066	81.6–61.2	13,055–9,791	115.3–86.5	18,447–13,835
1/2–20	57.0–42.7	9,114–5,835	91.9–69.0	14,711–11,033	130.0–97.4	20,787–15,590
9/16–12	73.0–54.7	10,374–7,780	117.7–88.1	16,744–12,558	166.4–124.8	23,660–17,745
9/16–18	81.4–61.0	11,571–8,678	131.3–98.1	18,676–14,007	185.6–139.2	26,390–19,793
5/8–11	100.6–75.5	12,882–9,662	162.4–121.8	20,792–15,594	229.5–172.1	29,380–22,035
5/8–18	114–85.5	14,592–10,944	184–138	23,552–17,664	260.0–195.0	33,280–24,960
3/4–10	178.5–133.9	19,038–14,279	288–216	30,728–23,046	407.1–305.3	43,420–35,368
3/4–16	199–149.5	21,261–15,946	321.7–241.3	34,316–25,737	454.6–341.0	48,490–45,045
7/8–9	288–216	26,334–19,751	464.9–348.7	42,504–31,878	656.9–492.7	60,060–45,045
7/8–14	317–238	29,013–19,751	512.2–384.1	46,828–35,121	723.7–542.8	66,170–49,628
1–8	432–324	34,542–25,907	696.9–522.7	55,752–41,814	984.8–738.6	78,780–59,085
1–12	472–354	37,791–28,343	761.1–571.8	60,996–45,747	1077–808	86,190–64,643

TABLE 9.1 Tightening Torque Requirements for American Standard Steel Bolts

CHAPTER 10
SPRING CALCULATIONS— DIE AND STANDARD TYPES

Springs and die springs are important mechanical components used in countless mechanisms, mechanical systems, and tooling applications. This chapter contains data and calculation procedures that are used to design springs and that also allow the machinist, toolmaker or tool engineer, metalworker, and designer to measure an existing spring and determine its spring rate. In most applications, normal spring materials are spring steel or music wire, while other applications require stainless steel, high-alloy steels, or beryllium-copper alloys. The main applications contained in this chapter apply to helical compression die springs and standard springs using round, square, and rectangular spring wire. Included are compression, extension, torsion, and flat or bowed spring equations used in design, specification, and replacement applications. Figure 10.1 shows some typical types of springs.

Material Selection. It is important to adhere to proper procedures and design considerations when designing springs.

Economy. Will economical materials such as ASTM A-229 wire suffice for the intended application?

Corrosion Resistance. If the spring is used in a corrosive environment, you may select materials such as 17-7 PH stainless steel or the other stainless steels, i.e., 301, 302, 303, 304, etc.

Electrical Conductivity. If you require the spring to carry an electric current, materials such as beryllium copper and phosphor bronze are available.

Temperature Range. Whereas low temperatures induced by weather are seldom a consideration, high-temperature applications call for materials such as 301 and 302 stainless steel, nickel-chrome A-286, 17-7 PH, Inconel 600, and Inconel X750. Design stresses should be as low as possible for springs designed for use at high operating temperatures.

Shock Loads, High Endurance Limit, and High Strength. Materials such as music wire, chrome-vanadium, chrome-silicon, 17-7 stainless steel, and beryllium copper are indicated for these applications.

General Spring Design Recommendations. Try to keep the ends of the spring, where possible, within such standard forms as closed loops, full loops to center, closed and ground, open loops, and so on.

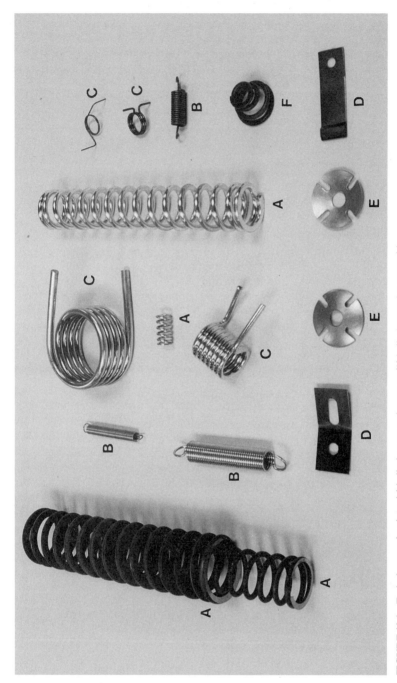

FIGURE 10.1 Typical types of springs: (*a*) helical compression types; (*b*) helical extension types; (*c*) torsion types; (*d*) flat springs, blue-steel and beryllium-copper types; (*e*) slotted spring washers; (*f*) conical compression type.

SPRING CALCULATIONS—DIE AND STANDARD TYPES　　10.3

Pitch. Keep the coil pitch constant unless you have a special requirement for a variable-pitch spring.

Keep the spring index D/d between 6.5 and 10 wherever possible. Stress problems occur when the index is too low, and entanglement and waste of material occur when the index is too high.

Do not electroplate the spring unless it is required by the design application. The spring will be subject to hydrogen embrittlement unless it is processed correctly after electroplating. Hydrogen embrittlement causes abrupt and unexpected spring failures. Plated springs must be baked at a specified temperature for a definite time interval immediately after electroplating to prevent hydrogen embrittlement. For cosmetic purposes and minimal corrosion protection, zinc electroplating is generally used, although other plating, such as chromium, cadmium, tin, etc., is also used according to the application requirements. Die springs usually come from the diespring manufacturers with colored enamel paint finishes for identification purposes. Black oxide and blueing are also used for spring finishes.

Special Processing Either During or After Manufacture. Shot peening improves surface qualities from the standpoint of reducing stress concentration points on the spring wire material. This process also can improve the endurance limit and maximum allowable stress on the spring. Subjecting the spring to a certain amount of permanent *set* during manufacture eliminates the set problem of high energy versus mass on springs that have been designed with stresses in excess of the recommended values. This practice is *not* recommended for springs that are used in critical applications.

Stress Considerations. Design the spring to stay within the allowable stress limit when the spring is fully compressed, or "bottomed." This can be done when there is sufficient space available in the mechanism and economy is not a consideration. When space is not available, design the spring so that its maximum working stress at its maximum working deflection does not exceed 40 to 45 percent of its minimum yield strength for compression and extension springs and 75 percent for torsion springs. Remember that the minimum yield strength allowable is different for differing wire diameters, the higher yield strengths being indicated for smaller wire diameters. See the later subsections for figures and tables indicating the minimum yield strengths for different wire sizes and different materials.

Direction of Winding on Helical Springs. Confusion sometimes exists as to what constitutes a right-hand or left-hand wound spring. Standard practice recognizes that the winding hand of helical springs is the same as standard right-hand screw thread and left-hand screw thread. A right-hand wound spring has its coils going in the same direction as a right-hand screw thread and the opposite for a left-hand spring. On a right-hand helical spring, the coil helix progresses away from your line of sight in a clockwise direction when viewed on end. This seems like a small problem, but it can be quite serious when designing torsion springs, where the direction of wind is critical to proper spring function. In a torsion spring, the coils must "close down" or tighten when the spring is deflected during normal operation, going back to its initial position when the load is removed. If a torsion spring is operated in the wrong direction, or "opened" as the load is applied, the working stresses become much higher and the spring could fail. The torsion spring coils also increase in diameter when operated in the wrong direction and likewise decrease in diameter when operated in the correct direction. See equations in Sec. 10.4.4 for calculations that show the final diameter of torsion springs when they are deflected during operation.

Also note that when two helical compression springs are placed one inside the other for a higher combined rate, the coil helixes must be wound opposite hand from each other. This prevents the coils from jambing or tangling during operation. Compression springs employed in this manner are said to be in *parallel*, with the final rate equal to the combined rate of the two springs added together. Springs that are employed one atop the other or in a straight line are said to be in *series*, with their final rate equal to 1 divided by the sum of the reciprocals of the separate spring rates.

EXAMPLE. Springs in parallel:

$$R_f = R_1 + R_2 + R_3 + \cdots + R_n$$

Springs in series:

$$\frac{1}{R_f} = \frac{1}{R_1} + \frac{1}{R_2} + \frac{1}{R_3} + \cdots + \frac{1}{R_n}$$

where R_f = final combined rate
$R_{1,2,3}$ = rate of each individual spring

In the following subsections you will find all the design equations, tables, and charts required to do the majority of spring work today. Special springs such as irregularly shaped flat springs and other nonstandard forms are calculated using the standard beam and column equations found in other handbooks, or they must be analyzed using involved stress calculations or prototypes made and tested for proper function.

Spring Design Procedures

1. Determine what spring rate and deflection or spring travel are required for your particular application.
2. Determine the space limitations the spring is required to work in, and try to design the spring accordingly using a parallel arrangement, if required, or allow space in the mechanism for the spring according to its calculated design dimensions.
3. Make a preliminary selection of the spring material dictated by the application or economics.
4. Make preliminary calculations to determine wire size or other stock size, mean diameter, number of coils, length, and so forth.
5. Perform the working stress calculations with the Wahl stress correction factor applied to see if the working stress is *below* the *allowable* stress.

The *working stress* is calculated using the appropriate equation with the working load applied to the spring. The load on the spring is found by multiplying the spring rate times the deflection length of the spring. For example, if the spring rate was calculated to be 25 lbf/in and the spring is deflected 0.5 in, then the load on the spring is 25 × 0.5 = 12.5 lbf.

The *maximum allowable stress* is found by multiplying the minimum tensile strength allowable for the particular wire diameter or size used in your spring times the appropriate multiplier. See the figures and tables in this chapter for minimum tensile strength allowables for different wire sizes and materials and the appropriate multipliers.

EXAMPLE. You are designing a compression spring using 0.130-in-diameter music wire, ASTM A-228. The allowable maximum stress for this wire size is

$$0.45 \times 258{,}000 = 116{,}100 \text{ psi} \quad \text{(see wire tables)}$$

NOTE. A more conservatively designed spring would use a multiplier of 40 percent (0.40), while a spring that is not cycled frequently can use a multiplier of 50 percent (0.50), with the spring possibly taking a slight set during repeated operations or cycles. The multiplier for torsion springs is 75 percent (0.75) in all cases and is conservative.

If the working stress in the spring is *below* the maximum allowable stress, the spring is properly designed relative to its stress level during operation. Remember that the modulus of elasticity of spring materials diminishes as the working temperature rises. This factor causes a decline in the spring rate. Also, working stresses should be decreased as the operating temperature rises. The figures and tables in this chapter show the maximum working temperature limits for different spring and spring wire materials. Only appropriate tests will determine to what extent these recommended limits may be altered.

10.1 HELICAL COMPRESSION SPRING CALCULATIONS

This section contains equations for calculating compression springs. Note that all equations throughout this chapter may be transposed for solving the required variable when all variables are known except one. The nomenclature for all symbols contained in the compression and extension spring design equations is listed in subsections of this chapter.

10.1.1 Round Wire

Rate:

$$R, \text{lb/in} = \frac{Gd^4}{8ND^3} \left.\vphantom{\frac{Gd^4}{8ND^3}}\right\} \text{Transpose for } d, N, \text{ or } D$$

Torsional stress:

$$S, \text{total corrected stress, psi} = \frac{8K_a DP}{\pi d^3} \left.\right\} \text{Transpose for } D, P, \text{ or } d$$

Wahl curvature-stress correction factor:

$$K_a = \frac{4C-1}{4C-4} + \frac{0.615}{C} \qquad \text{where } C = \frac{D}{d}$$

10.1.2 Square Wire

Rate:

$$R, \text{lb/in} = \frac{Gt^4}{5.6ND^3} \left.\right\} \text{Transpose for } t, N, \text{ or } D$$

Torsional stress:

$$S, \text{total corrected stress, psi} = \frac{2.4K_{a1}DP}{t^3} \left.\right\} \text{Transpose for } D, P, \text{ or } t$$

Wahl curvature-stress correction factor:

$$K_{a1} = 1 + \frac{1.2}{C} + \frac{0.56}{C^2} + \frac{0.5}{C^3} \qquad \text{where } C = \frac{D}{t}$$

10.1.3 Rectangular Wire

Rate (see Fig. 10.2 for a table of factors K_1 and K_2):

$$R, \text{lb/in} = \frac{Gbt^3}{ND^3} K_2 \left.\right\} \text{Transpose for } b, t, N, \text{ or } D$$

Torsional stress, corrected:

$$S, \text{psi} = \frac{PD}{bt\sqrt{bt}} \beta \left.\right\} \text{Transpose for } b, t, P, \text{ or } D$$

NOTE. β is obtained from Fig. 10.2.

10.1.4 Solid Height of Compression Springs

For round wire, see Fig. 10.3.

For Square and Rectangular Wire. Due to distortion of the cross section of square and rectangular wire when the spring is formed, the compressed solid height can be determined from

SPRING CALCULATIONS—DIE AND STANDARD TYPES

TABLE FACTORS FOR SQUARE AND RECTANGULAR SECTIONS

b · t	1	1.2	1.5	2	2.5	3	5	10	∞
Factor K_1	0.416	0.438	0.462	0.492	0.516	0.534	0.582	0.624	0.666
Factor K_2	0.180	0.212	0.250	0.292	0.317	0.335	0.371	0.398	0.424

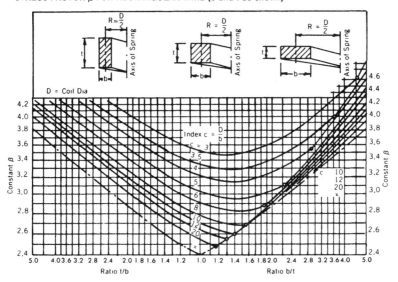

FIGURE 10.2 Stress factors for rectangular wire and K factors.

	Type of End			
	Open or Plain (not ground)	Open or Plain (with ends ground)	Squared or Closed (not ground)	Closed and Ground
Feature	Formula			
Pitch (p)	$\dfrac{FL - d}{N}$	$\dfrac{FL}{TC}$	$\dfrac{FL - 3d}{N}$	$\dfrac{FL - 2d}{N}$
Solid Height (SH)	(TC + 1)d	TC × d	(TC + 1)d	TC × d
Number of Active Coils (N)	N = TC or $\dfrac{FL - d}{p}$	N = TC − 1 or $\dfrac{FL}{p} - 1$	N = TC − 2 or $\dfrac{FL - 3d}{p}$	N = TC − 2 or $\dfrac{FL - 2d}{p}$
Total Coils (TC)	$\dfrac{FL - d}{p}$	$\dfrac{FL}{p}$	$\dfrac{FL - 3d}{p} + 2$	$\dfrac{FL - 2d}{p} + 2$
Free Length (FL)	(p × TC) + d	p × TC	(p × N) + 3d	(p × N) + 2d

d = wire dia.

FIGURE 10.3 Compression-spring features.

$$t' = 0.48t \left(\frac{OD}{D} + 1 \right)$$

where t' = new thickness of inner edge of section in the axial direction, after coiling
  t = thickness of section before coiling
  D = mean diameter of the spring
  OD = outside diameter

Active Coils in Compression Springs. Style of ends may be selected as follows:

- *Open ends, not ground.* All coils are active.
- *Open ends, ground.* One coil is inactive.
- *Closed ends, not ground.* Two coils are inactive.
- *Closed ends, ground.* Two coils are inactive.

When using the compression spring equations, the variable N refers to the number of *active* coils in the spring being calculated.

10.2 HELICAL EXTENSION SPRINGS (CLOSE-WOUND)

This type of spring is calculated using the same equations for the standard helical compression spring, namely, rate, stress, and Wahl stress-correction factor. One exception when working with helical extension springs is that this type of spring is sometimes wound by the spring manufacturer with an initial tension in the wire. This initial tension keeps the coils tightly closed together and creates a pretension in the spring. When designing the spring, you may specify the initial tension on the spring, in pounds. When you do specify the initial tension, you must calculate the torsional stress developed in the spring as a result of this initial tension.

First, calculate torsional stress S_i due to initial tension P_1 in

$$S_i = \frac{8DP_1}{\pi d^3}$$

where P_1 = initial tension, lb. Second, for the value of S_i calculated and the known spring index D/d, determine on the graph in Fig. 10.4 whether or not S_i appears in the preferred (shaded) area. If S_i falls in the shaded area, the spring can be produced readily. If S_i is above the shaded area, reduce it by increasing the wire size. If S_i is below the shaded area, select a smaller wire size. In either case, recalculate the stress and alter the number of coils, axial space, and initial tension as necessary.

10.3 SPRING ENERGY CONTENT OF COMPRESSION AND EXTENSION SPRINGS

The potential energy which may be stored in a deflected compression or extension spring is given by

$$P_e = \frac{Rs^2}{2}$$

SPRING CALCULATIONS—DIE AND STANDARD TYPES

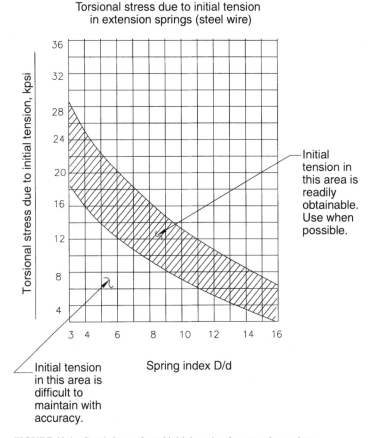

FIGURE 10.4 Graph for preferred initial tension for extension springs.

Also:

$$P_e = \frac{1}{2} R(s_2^2 - s_1^2) \quad \text{in moving from point } s_1 \text{ to } s_2$$

where R = rate of the spring, lb/in, lb/ft, N/m
 s = distance spring is compressed or extended, in, m
 P_e = potential energy, in · lb, ft · lb, J
 s_1, s_2 = distances moved, in

EXAMPLE. A compression spring with a rate of 50 lb/in is compressed 4 in. What is the potential energy stored in the loaded spring?

$$P_e = \frac{50(4)^2}{2} = 400 \text{ in} \cdot \text{lb or } 33.33 \text{ ft} \cdot \text{lb}$$

Thus the spring will perform 33.33 ft · lb of work energy when released from its loaded position. Internal losses are negligible. This procedure is useful to mechanical designers and tool engineers who need to know the work a spring will produce in a mechanism or die set and the input energy required to load the spring.

Expansion of Compression Springs When Deflected. A compression spring outside diameter will expand when the spring is compressed. This may pose a problem if the spring must work within a tube or cylinder and its outside diameter is close to the inside diameter of the containment. The following equation may be used to calculate the amount of expansion that takes place when the spring is compressed to solid height. For intermediate heights, use the percent of compression multiplied by the total expansion.

Total expansion = outside diameter (solid) − outside diameter

Expanded diameter is

$$\text{Outside diameter, solid} = \sqrt{D^2 + \frac{p^2 - d^2}{\pi^2}} + d$$

where p = pitch (distance between adjacent coil center lines), in
d = wire diameter, in
D = mean diameter of the spring, in

and outside diameter, solid = expanded diameter when compressed solid, in

Symbols for Compression and Extension Springs

R = rate, pounds of load per inch of deflection
P = load, lb
F = deflection, in
D = mean coil diameter, OD − d
d = wire diameter, in
t = side of square wire or thickness of rectangular wire, in
b = width of rectangular wire, in
G = torsional modulus of elasticity, psi
N = number of active coils, determined by the types of ends on a compression spring; equal to *all* the coils of an extension spring
S = torsional stress, psi
OD = outside diameter of coils, in
ID = inside diameter, in
C = spring index D/d
L = length of spring, in
H = solid height, in
K_a = Wahl stress-correction factor
K_1, K_2, β (see Fig. 10.2)

For preferred and special end designs for extension springs, see Fig. 10.5.

SPRING CALCULATIONS—DIE AND STANDARD TYPES 10.11

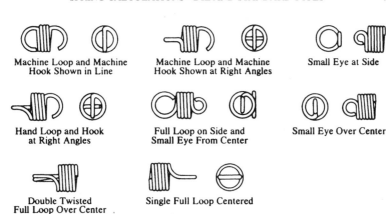

ALL THE ABOVE ENDS ARE STANDARD TYPES FOR WHICH
NO SPECIAL TOOLS ARE REQUIRED

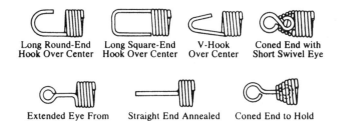

THIS GROUP OF SPECIAL ENDS REQUIRES SPECIAL TOOLS

FIGURE 10.5 Preferred and special ends, extension springs.

10.4 TORSION SPRINGS

Refer to Fig. 10.6.

10.4.1 Round Wire

Moment (torque) is

$$M, \text{lb} \cdot \text{in} = \frac{Ed^4 T}{10.8ND} \quad \Big\} \text{Transpose for } d, T, N, \text{ or } D$$

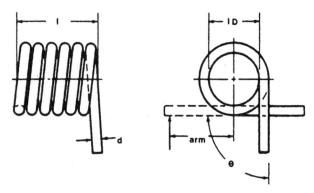

FIGURE 10.6 Torsion spring.

Tensile stress is

$$S, \text{psi} = \frac{32M}{\pi d^3} K \Bigg\} \text{Transpose for } M \text{ or } d$$

10.4.2 Square Wire

Moment (torque) is

$$M, \text{lb} \cdot \text{in} = \frac{Ed^4 T}{6.6ND} \Bigg\} \text{Transpose for } t, T, N, \text{ or } D$$

Tensile stress is

$$S, \text{psi} = \frac{6M}{t^3} K_1 \Bigg\} \text{Transpose for } M \text{ or } t$$

The stress-correction factor K or K_1 for torsion springs with round or square wire, respectively, is applied according to the spring index as follows:

$$\begin{aligned} \text{When spring index} &= 6, \quad K = 1.15 \\ &= 8, \quad K = 1.11 \\ &= 10, \quad K = 1.08 \end{aligned} \Bigg\} \text{for round wire}$$

$$\begin{aligned} \text{When spring index} &= 6, \quad K_1 = 1.13 \\ &= 8, \quad K_1 = 1.09 \\ &= 10, \quad K_1 = 1.07 \end{aligned} \Bigg\} \text{for square wire}$$

For spring indexes that fall between the values shown, interpolate the new correction factor value. Use standard interpolation procedures.

SPRING CALCULATIONS—DIE AND STANDARD TYPES 10.13

10.4.3 Rectangular Wire

Moment (torque) is

$$M, \text{lb} \cdot \text{in} = \frac{Ebt^3T}{6.6ND} \bigg\} \text{Transpose for } b, t, T, N, \text{ or } D$$

Tensile stress is

$$S, \text{psi} = \frac{6M}{bt^2} \bigg\} \text{Transpose for } M, t, \text{ or } b$$

10.4.4 Symbols, Diameter Reduction, and Energy Content

Symbols for Torsion Springs

D = mean coil diameter, in
d = diameter of round wire, in
N = total number of coils, i.e., 6 turns, 7.5 turns, etc.
E = torsional modulus of elasticity (see charts in this chapter)
T = revolutions through which the spring works (e.g., 90° arc = 90/360 = 0.25 revolutions, etc.)
S = bending stress, psi
M = moment or torque, lb · in
b = width of rectangular wire, in
t = thickness of rectangular wire, in
K, K_1 = stress-correction factor for round and square wire, respectively

Torsion Spring Reduction of Diameter During Deflection. When a torsion spring is operated in the correct direction (coils close down when load is applied), the spring's inside diameter (ID) is reduced as a function of the number of degrees the spring is rotated in the closing direction and the number of coils. This may be calculated from the following equation:

$$ID_r = \frac{360N(ID_f)}{360N + R°}$$

where ID_r = inside diameter after deflection (closing), in
ID_f = inside diameter before deflection (free), in
N = number of coils
$R°$ = number of degrees rotated in the closing direction

NOTE. When a spring is manufactured, great care must be taken to ensure that *no* marks or indentations are formed on the spring coils.

Spring Energy Content (Torsion, Coil, or Spiral Springs). In the case of a torsion or spiral spring, the potential energy P_e the spring will contain when deflected in the closing direction can be calculated from

$$P_e = \frac{1}{2} R\theta_r^2 \quad \text{also} \quad M = R\theta_r$$

where M = resisting torque, lb · ft, N · m
R = spring rate, lb/rad, N/rad
θ_r = angle of deflection, rad

Remember that 2π rad = 360° and 1 rad = 0.01745°.

NOTE. Units of elastic potential energy are the same as those for work and are expressed in foot pounds in the U.S. customary system and in joules in SI. Although spring rates for most commercial springs are not strictly linear, they are close enough for most calculations where extreme accuracy is not required.

In a similar manner, the potential energy content of leaf and beam springs can be derived approximately by finding the apparent rate and the distance through which the spring moves.

Symbols for Spiral Torsion Springs (and Flat Springs,* Sec. 10.5)

*E = bending modulus of elasticity, psi (e.g., 30×10^6 for most steels)
θ_r = angular deflection, rad (for energy equations)
θ = angular deflection, revolutions (e.g., 90° = 0.25 revolutions)
*L = length of active spring material, in
M = moment or torque, lb · in
*b = material width, in
*t = material thickness, in
A = arbor diameter, in
OD_f = outside diameter in the free condition

10.5 FLAT SPRINGS

Cantilever Spring. Load (see Figs. 10.7a, b, and c) is

$$P, \text{lb} = \frac{EFbt^3}{4L^3} \left.\right\} \text{Transpose for } F, b, t, \text{ or } L$$

Stress is

$$S, \text{psi} = \frac{3EFt}{2L^2} = \frac{6PL}{bt^2} \left.\right\} \text{Transpose for } F, t, L, b, \text{ or } P$$

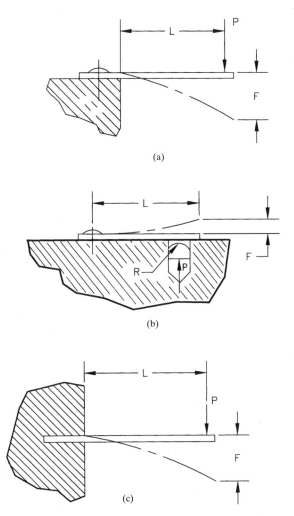

FIGURE 10.7 Flat springs, cantilever.

Simple Beam Springs.   Load (see Figs. 10.8a and b) is

$$P, \text{lb} = \frac{4EFbt^3}{L^3} \bigg\} \text{Transpose for } F, b, t, \text{ or } L$$

Stress is

$$S, \text{psi} = \frac{6EFt}{L^2} = \frac{3PL}{2bt^2} \bigg\} \text{Transpose for } F, b, t, L, \text{ or } P$$

In highly stressed spring designs, the spring manufacturer should be consulted and its recommendations followed. Whenever possible in mechanism design, space

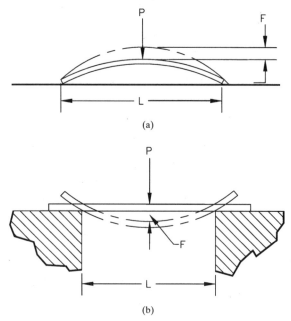

FIGURE 10.8 Flat springs, beam.

for a moderately stressed spring should be allowed. This will avoid the problem of marginally designed springs, that is, springs that tend to be stressed close to or beyond the maximum allowable stress. This, of course, is not always possible, and adequate space for moderately stressed springs is not always available. Music wire and some of the other high-stress wire materials are commonly used when high stress is a factor in design and cannot be avoided.

10.6 SPRING MATERIALS AND PROPERTIES

See Fig. 10.9 for physical properties of spring wire and strip that are used for spring design calculations.

Minimum Yield Strength for Spring-Wire Materials.   See Fig. 10.10 for minimum yield strengths of spring-wire materials in various diameters: (*a*) stainless steels, (*b*) chrome-silicon/chrome vanadium alloys, (*c*) copper-base alloys, (*d*) nickel-base alloys, and (*e*) ferrous.

Buckling of Unsupported Helical Compression Springs.   Unsupported or unguided helical compression springs become unstable in relation to their slenderness ratio and deflection percentage of their free length. Figure 10.11 may be used to determine the unstable condition of any particular helical compression spring under a particular deflection load or percent of free length.

SPRING CALCULATIONS—DIE AND STANDARD TYPES   10.17

Material and specification	E, 10^6 psi	G, 10^6 psi	Design stress, % min. yield	Conductivity, % IACS	Density, lb/in^3	Max. operating temperature, °F	FA*	SA*
High-carbon steel wire								
Music ASTM A228	30	11.5	45	7	0.284	250	E	H
Hard-drawn								
ASTM A227	30	11.5	40	7	0.284	250	P	M
ASTM A679	30	11.5	45	7	0.284	250	P	M
Oil-tempered								
ASTM A229	30	11.5	45	7	0.284	300	P	M
Carbon valve	30	11.5						
ASTM A230	30	11.5	45	7	0.284	300	E	H
Alloy steel wire								
Chrome-vanadium ASTM A231	30	11.5	45	7	0.284	425	E	H
Chrome-silicon ASTM A401	30	11.5	45	5	0.284	475	F	H
Silicon-manganese AISI 9260	30	11.5	45	4.5	0.284	450	F	H
Stainless steel wire								
AISI 302/304 ASTM A313	28	10	35	2	0.286	550	G	M
AISI 316 ASTM A313	28	10	40	2	0.286	550	G	M
17-7PH ASTM A313(631)	29.5	11	45	2	0.286	650	G	H
Nonferrous alloy wire								
Phosphor-bronze ASTM B159	15	6.25	40	18	0.320	200	G	M
Beryllium-copper ASTM B197	18.5	7	45	21	0.297	400	E	H
Monel 400 AMS 7233	26	9.5	40	—	—	450	F	M
Monel K500 QQ-N-286	26	9.5	40	—	—	550	F	M

FIGURE 10.9   Spring materials and properties.

Material and specification	E, 10^6 psi	G, 10^6 psi	Design stress, % min. yield	Conductivity, % IACS	Density, lb/in^3	Max. operating temperature, °F	FA*	SA*
High-temperature alloy wire									
Nickel-chrome ASTM A286	29	10.4	35	2	0.290	510	—	L
Inconel 600 QQ-W-390	31	11	40	1.5	0.307	700	F	L
Inconel X750 AMS 5698, 5699	31	12	40	1	0.298	1100	F	L
High-carbon steel strip									
AISI 1065	30	11.5	75	7	0.284	200	F	M
AISI 1075	30	11.5	75	7	0.284	250	G	H
AISI 1095	30	11.5	75	7	0.284	250	E	H
Stainless steel strip									
AISI 301	28	10.5	75	2	0.286	300	G	M
AISI 302	28	10.5	75	2	0.286	550	G	M
AISI 316	28	10.5	75	2	0.286	550	G	M
17-7PH ASTM A693	29	11	75	2	0.286	650	G	H
Nonferrous alloy strip									
Phosphor-bronze ASTM B103	15	6.3	75	18	0.320	200	G	M
Beryllium-copper ASTM B194	18.5	7	75	21	0.297	400	E	H
Monnel 400 AMS 4544	26	—	75	—	—	450	—	—
Monel K500 QQ-N-286	26	—	75	—	—	550	—	—
High-temperature alloy strip									
Nickel-chrome ASTM A286	29	10.4	75	2	0.290	510	—	L
Inconel 600 ASTM B168	31	11	40	1.5	0.307	700	F	L
Inconel X750 AMS 5542	31	12	40	1	0.298	1100	F	L

*Letter designations of the last two columns indicate: FA = fatigue applications; SA = strength applications; E = excellent; G = good; F = fair; L = low; H = high; M = medium; P = poor.

FIGURE 10.9 (*Continued*) Spring materials and properties.

Stainless steels

Wire size, in	Type 302	Type 17-7 PH*	Wire size, in	Type 302	Type 17-7 PH*	Wire size, in	Type 302	Type 17-7 PH*
0.008	325	345	0.033	276		0.060	256	
0.009	325		0.034	275		0.061	255	305
0.010	320	345	0.035	274		0.062	255	297
0.011	318	340	0.036	273		0.063	254	
0.012	316		0.037	272		0.065	254	
0.013	314		0.038	271		0.066	250	
0.014	312		0.039	270		0.071	250	297
0.015	310	340	0.040	270		0.072	250	292
0.016	308	335	0.041	269	320	0.075	250	
0.017	306		0.042	268	310	0.076	245	
0.018	304		0.043	267		0.080	245	292
0.019	302		0.044	266		0.092	240	279
0.020	300	335	0.045	264		0.105	232	274
0.021	298	330	0.046	263		0.120		272
0.022	296		0.047	262		0.125		272
0.023	294		0.048	262		0.131		260
0.024	292		0.049	261		0.148	210	256
0.025	290	330	0.051	261	310	0.162	205	256
0.026	289	325	0.052	260	305	0.177	195	
0.027	267		0.055	260		0.192		
0.028	266		0.056	259		0.207	185	
0.029	284		0.057	258		0.225	180	
0.030	282	325	0.058	258		0.250	175	
0.031	280	320	0.059	257		0.375	140	
0.032	277							

FIGURE 10.10a Stainless steel wire.

Chrome-silicon/chrome-vanadium steels

Wire size, in	Chrome-silicon	Chrome-vanadium
0.020		300
0.032	300	290
0.041	298	280
0.054	292	270
0.062	290	265
0.080	285	255
0.092	280	
0.105		245
0.120	275	
0.135	270	235
0.162	265	225
0.177	260	
0.192	260	220
0.218	255	
0.250	250	210
0.312	245	203
0.375	240	200
0.437		195
0.500		190

FIGURE 10.10b Chrome silicon/chrome vanadium.

Copper-base alloys

Wire size range, 1 in	Strength
Phosphor-bronze (grade A)	
0.007–0.025	145
0.026–0.062	135
0.063 and over	130
Beryllium-copper (alloy 25 pretempered)	
0.005–0.040	180
0.041 and over	170
Spring brass (all sizes)	120

FIGURE 10.10c Copper-base alloys.

Nickel-base alloys

Inconel (spring temper)

Wire size range, 1 in	Strength
Up to 0.057	185
0.057–0.114	175
0.114–0.318	170
Inconel X (spring temper)*	190–220

FIGURE 10.10d Nickel-base alloys.

Ferrous

Wire size, in	Music wire	Hard drawn	Oil temp.	Wire size, in	Music wire	Hard drawn	Oil temp.	Wire Size, in	Music wire	Hard drawn	Oil temp.
0.006	399	307	315	0.046	309	249	259	0.094	274	219	
0.009	393	305	313	0.047	309	248		0.095	274		
0.010	387	303	311	0.048	306	247		0.099	274		
0.011	382	301	309	0.049	306	246		0.100	271		
0.012	377	299	307	0.050	306	245		0.101	271		
0.013	373	297	305	0.051	303	244		0.102	270		
0.014	369	295	303	0.052	303	244		0.105	270	216	225
0.015	365	293	301	0.053	303	243	253	0.106	268		
0.016	362	291	300	0.054	303	243		0.109	268		
0.017	362	289	298	0.055	300	242		0.110	267		
0.018	356	287	297	0.056	300	241		0.111	267		
0.019	356	285	295	0.057	300	240		0.112	266		
0.020	350	283	293	0.058	300	240		0.119	266		
0.021	350	281		0.059	296	239		0.120	263	210	220
0.022	345	280		0.060	296	238	247	0.123	263		
0.023	345	278	289	0.061	296	237		0.124	261		
0.024	341	277		0.062	296	237		0.129	261		
0.025	341	275	286	0.063	293	236		0.130	258		
0.026	337	274		0.064	293	235		0.135	258	206	215
0.027	337	272		0.065	293	235		0.139	258		
0.028	333	271	283	0.066	290			0.140	256		
0.029	333	267		0.067	290	234		0.144	256		
0.030	330	266		0.069	290	233		0.145	254	203	210
0.031	330	266	280	0.070	289			0.148	254		

FIGURE 10.10e Ferrous spring wire.

Wire size, in	Music wire	Hard drawn	Oil temp.	Wire size, in	Music wire	Hard drawn	Oil temp.	Wire Size, in	Music wire	Hard drawn	Oil temp.
0.032	327	265		0.071	288			0.149	253		
0.033	327	264		0.072	287	232	241	0.150	253		
0.034	324	262	274	0.074	287	231		0.151	251		
0.035	324	261		0.075	287			0.160	251		
0.036	321	260		0.076	284	230		0.161	249		
0.037	321	258		0.078	284	229		0.162	249	200	205
0.038	318	257		0.079	284		235	0.177	245	195	200
0.039	318	256		0.080	282	227		0.192	241	192	195
0.040	315	255		0.083	282			0.207	238	190	190
0.041	315	255	266	0.084	279	225		0.225	235	186	188
0.042	313	254		0.085	279			0.250	230	182	185
0.043	313	252		0.089	279			0.3125		174	183
0.044	313	251		0.090	276	222	230	0.375		167	180
0.045	309	250		0.091	276			0.4375		165	175
				0.092	276			0.500		156	170
				0.093	276						

Note: Values in table are psi × 10^3.
* After aging.

FIGURE 10.10e *(Continued)* Ferrous spring wire.

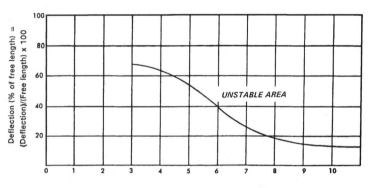

FIGURE 10.11 Buckling of helical compression springs.

10.7 ELASTOMER SPRINGS

Elastomer springs have proven to be the safest, most efficient, and most reliable compression material for use with punching, stamping, and drawing dies and blank-holding and stripper plates. These springs feature no maintenance and very long life, coupled with higher loads and increased durability. Other stock sizes are available than those shown in Tables 10.1 and 10.2. Elastomer springs are used where metallic springs cannot be used, i.e., in situations requiring chemical resistance, nonmagnetic properties, long life, or other special properties. See Fig. 10.12 for dimensional reference to Tables 10.1 and 10.2.

TABLE 10.1 Elastomer Springs (Standard)

D, in	d, in	L, in	R*	Deflection[†]	T[‡]
0.630	0.25	0.625	353	0.22	77
0.630	0.25	1.000	236	0.34	83
0.787	0.33	0.625	610	0.22	133
0.787	0.33	1.000	381	0.35	133
1.000	0.41	1.000	598	0.35	209
1.000	0.41	1.250	524	0.44	229
1.250	0.53	1.250	1030	0.44	451
1.250	0.53	2.500	517	0.87	452
1.560	0.53	1.250	1790	0.44	783
1.560	0.53	2.500	930	0.87	815
2.000	0.66	2.500	1480	0.87	1297
2.500	0.66	2.500	2286	0.87	2000
3.150	0.83	2.500	4572	0.87	4000

See Fig. 10.11 for dimensions D, d, and L.
* Spring rate, lb/in, ±20%.
[†] Maximum deflection = 35% of L.
[‡] Approximate total load at maximum deflection ±20%.
Source: Reid Tool Supply Company, Muskegon, MI 49444-2684.

SPRING CALCULATIONS—DIE AND STANDARD TYPES   10.23

TABLE 10.2 Urethane Springs (95 Durometer, Shore A Scale)

D, in	d, in	L, in	Load, lb, ⅛-in deflection
0.875	0.250	1.000	425
0.875	0.250	1.250	325
0.875	0.250	1.750	250
1.000	0.375	1.000	525
1.500	0.375	1.500	325
1.125	0.500	1.000	600
1.125	0.500	2.000	275
1.250	0.625	1.000	700
1.250	0.625	2.000	325
1.500	0.750	1.250	875
1.500	0.750	2.000	525
2.000	1.000	1.250	1550
2.000	1.000	2.750	625

See Fig. 10.12 for dimensions D, d, and L.
Temperature range: –40°F to +180°F, color black.
Source: Reid Tool Supply Company, Muskegon, MI 49444-2684.

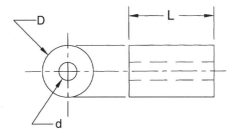

FIGURE 10.12   Dimensional reference to Tables 10.1 and 10.2.

10.8 BENDING AND TORSIONAL STRESSES IN ENDS OF EXTENSION SPRINGS

Bending and torsional stresses develop at the bends in the ends of an extension spring when the spring is stretched under load. These stresses should be checked by the spring designer after the spring has been designed and dimensioned. Alterations to the ends and radii may be required to bring the stresses into their allowable range (see Sec. 10.5 and Fig. 10.13).

The bending stress may be calculated from

$$\text{Bending stress at point } A = S_b = \frac{16PD}{\pi d^3} \left(\frac{r_1}{r_2} \right)$$

The torsional stress may be calculated from

$$\text{Torsional stress at point } B = S_t = \frac{8PD}{\pi d^3} \left(\frac{r_3}{r_4} \right)$$

Check the allowable stresses for each particular wire size of the spring being calculated from the wire tables. The calculated bending and torsional stresses cannot exceed the allowable stresses for each particular wire size. As a safety precaution, take 75 percent of the allowable stress shown in the tables as the minimum allowable when using the preceding equations.

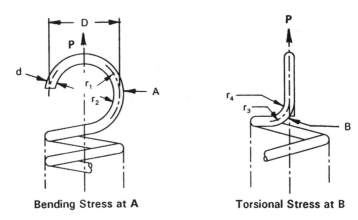

FIGURE 10.13 Bending and torsional stresses at ends of extension springs.

10.9 SPECIFYING SPRINGS, SPRING DRAWINGS, AND TYPICAL PROBLEMS AND SOLUTIONS

When a standard spring or a die spring collapses or breaks in operation, the reasons are usually as indicated by the following causes:

- Defective spring material
- Incorrect material for the application
- Spring cycled beyond its normal life
- Defect in manufacture such as nicks, notches, and deep forming marks on spring surface
- Spring incorrectly designed and overstressed beyond maximum allowable level
- Hydrogen embrittlement due to plating and poor processing (no postbaking used)
- Incorrect heat treatment

Specifying Springs and Spring Drawings. The correct dimensions must be specified to the spring manufacturer. See Figs. 10.14a, b, and c for dimensioning compression, extension, and torsion springs.

A typical engineering drawing for specifying a compression spring is shown in Fig. 10.15. Extension and torsion springs are also specified with a drawing similar to that shown in Fig. 10.15, using Figs. 10.14a, b, and c as a guide.

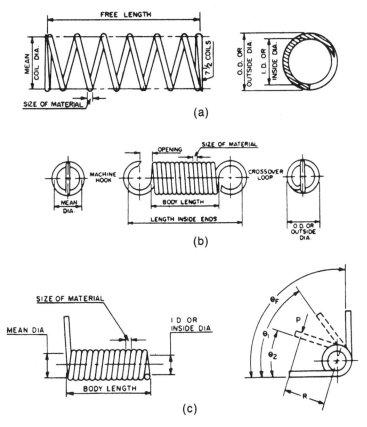

FIGURE 10.14 Dimensions required for springs: (*a*) compression springs; (*b*) extension springs; (*c*) torsion springs.

Typical Spring Problems and Solutions

Problem. A compression type die spring, using square wire, broke during use, and the original specification drawing is not available.

Solution. Measure the outside diameter, inside diameter, cross section or diameter of wire, free length of spring, number of coils or turns, and the distance the spring was deflected in operation. Remember, if a compression spring has closed and ground ends (which die springs usually have), count the total number of coils or turns and subtract 2 coils to find the number of *active coils*. See Fig. 10.3 for the number of active coils for each type of end on compression springs. Most die springs use hard-drawn, oil-tempered, or valve spring material (see Fig. 10.9 for material specifications).

Then, use the appropriate minimum stress allowable for the spring's measured wire size, as shown in Fig. 10.9a. Stress levels in these figures represent thousands of pounds per square inch (i.e., if the charted value is 325, then the allowable minimum

10.26 CHAPTER TEN

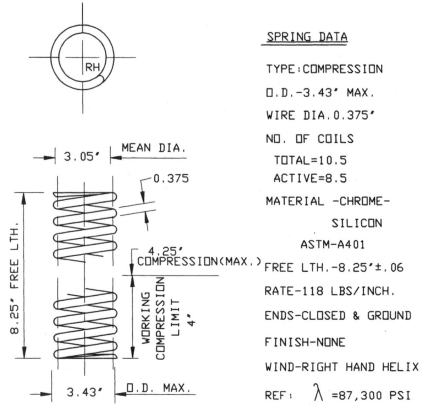

FIGURE 10.15 Typical engineering drawing for use by spring manufacturers.

tensile stress is 325,000 psi). Multiply this value by the appropriate correct stress allowable for compression springs, which is 45 percent or $0.45 \times 325{,}000 = 146{,}250$ psi.

With the preceding data and measurements, calculate the spring rate and the maximum stress the spring was subjected to during operation using the following procedure.

Step 1. See the equations shown in Sec. 10.1 for your application (round, square, or rectangular wire).

Step 2. Calculate the spring rate R.

Step 3. Calculate the working stress (torsional stress S) to see if it is within the allowable stress as indicated previously. If the stress level calculated for the broken spring is higher than the maximum allowable stress, select a material such as chrome-silicon or chrome-vanadium steel.

Step 4. If the calculated working stress level is below the maximum allowable, the spring may be ordered with all the dimensions and spring rate provided to the spring manufacturer.

SPRING CALCULATIONS—DIE AND STANDARD TYPES 10.27

$G := 11500000 \quad d := 0.250 \quad D := 1.700 \quad N := 13 \quad \dfrac{D}{d} = 6.8 \text{ Index C}$

$C := 6.8 \quad P := 250, 260 \,..\, 400 \quad K := 1.22 \text{ Wahl stress correction factor}$

$\dfrac{4 \cdot C - 1}{4 \cdot C - 4} + \dfrac{0.615}{C} = 1.22 \quad \dfrac{G \cdot d^4}{8 \cdot N \cdot D^3} = 87.918 \quad \text{RATE} = 87.92 \text{ lb/in}$

$\dfrac{8 \cdot K \cdot D \cdot P}{\pi \cdot d^3} = \text{STRESS, psi}$

8.45 · 10⁴	By assigning a range variable to P, which is the load on the spring, Math-Cad 7 will present a table of stress values from which the maximum allowable stress can be determined for a particular load P. In this problem, the maximum stress is indicated in the table as 118,300 psi, when the spring is loaded to 350 lbf. Maximum tensile strength for 0.250 diameter music wire (ASTM A-228) is 0.50 × 230,000 = 115,000 psi, which is close to the value in the table for the 350 lbf load. The spring is stressed slightly above the allowable of 50% of maximum tensile strength for the wire diameter indicated in the problem. This proved to be adequate design for this particular spring, which was cycled infrequently in operation. Operating temperature range for this application was from –40 to 150°F. Approximately 90,000 springs were used over a time span of 15 years without any spring failures.
8.788 · 10⁴	
9.126 · 10⁴	
9.464 · 10⁴	
9.802 · 10⁴	
1.014 · 10⁵	
1.048 · 10⁵	
1.082 · 10⁵	
1.115 · 10⁵	
1.149 · 10⁵	
* 1.183 · 10⁵	
1.217 · 10⁵	
1.251 · 10⁵	
1.284 · 10⁵	
1.318 · 10⁵	
1.352 · 10⁵	* Maximum stress level, psi, when the load is 350 lbf.

FIGURE 10.16 Compression spring calculation using MathCad PC program.

NOTE. Figure 10.15 shows a typical engineering drawing for ordering springs from the spring manufacturer, and Fig. 10.16 shows a typical compression spring calculation procedure.

CHAPTER 11
MECHANISMS, LINKAGE GEOMETRY, AND CALCULATIONS

The mechanisms and linkages discussed in this chapter have many applications for the product designer, tool engineer, and others involved in the design and manufacture of machinery, tooling, and mechanical devices and assemblies used in the industrial context. A number of important mechanical linkages are shown in Sec. 11.5, together with the mathematical calculations that govern their operation.

Mechanisms and Principles of Operation. When you study the operating principles of these devices, you will be able to see the relationships they have with the basic simple machines such as the lever, wheel and axle, inclined plane or wedge, gear wheel, and so forth. There are seven basic simple machines from which all machines and mechanisms may be constructed either singly or in combination, including the Rolomite mechanism. The hydraulic cylinder and gear wheel are also considered members of the basic simple machines.

Shown in Sec. 11.4 are other mechanisms which are used for tool-clamping purposes.

A number of practical mechanisms are shown in Sec. 11.3 together with explanations of their operation, in terms of their operational equations.

11.1 MATHEMATICS OF THE EXTERNAL GENEVA MECHANISM

See Figs. 11.1 and 11.2.

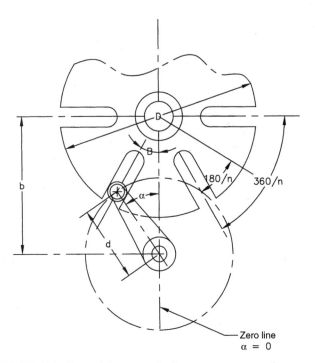

FIGURE 11.1 External Geneva mechanism.

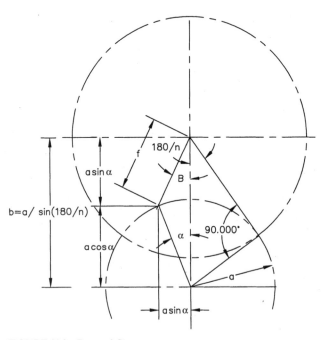

FIGURE 11.2 External Geneva geometry.

Kinematics of the External Geneva Drive. Assumed or given: $a, n, d,$ and p.

a = crank radius of driving member   and $m = \dfrac{1}{\sin(180/n)}$

n = number of slots in drive

d = roller diameter

p = constant velocity of driving crank, rpm

b = center distance = am

D = diameter of driven Geneva wheel = $2\sqrt{\dfrac{d^2}{4} + a^2 \cot^2 \dfrac{180}{n}}$

ω = constant angular velocity of driving crank = $p\pi/30$ rad/sec

α = angular position of driving crank at any time

β = angular displacement of driven member corresponding to crank angle α.

$$\cos \beta = \dfrac{m - \cos \alpha}{\sqrt{1 + m^2 - 2m \cos \alpha}}$$

Angular velocity of driven member = $\dfrac{d\beta}{dt} = \omega\left(\dfrac{m \cos \alpha - 1}{1 + m^2 - 2m \cos \alpha}\right)$

Angular acceleration of driven member = $\dfrac{d^2\beta}{dt^2} = \omega^2\left(\dfrac{m \sin \alpha (1 - m^2)}{(1 + m^2 - 2m \cos \alpha)^2}\right)$

Maximum angular acceleration occurs when $\cos \alpha = \sqrt{\left(\dfrac{1 + m^2}{4m}\right)^2 + 2} - \left(\dfrac{1 + m^2}{4m}\right)$

Maximum angular velocity occurs at $\alpha = 0°$ and equals $\dfrac{\omega}{m-1}$ rad/sec

11.2 MATHEMATICS OF THE INTERNAL GENEVA MECHANISM

See Figs. 11.3 and 11.4.

Equations for the Internal Geneva Wheel. Assumed or given: $a, n, d,$ and p.

a = crank radius of driving member   and $m = \dfrac{1}{\sin(180/n)}$

n = number of slots

d = roller diameter

p = constant velocity of driving crank, rpm

b = center distance = am

D = inside diameter of driven member = $2\sqrt{\dfrac{d^2}{4} + a^2 \cot^2 \dfrac{180}{n}}$

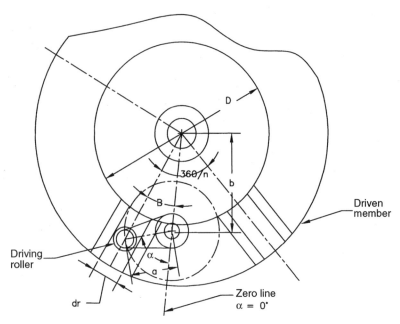

FIGURE 11.3 Internal Geneva mechanism (six-slot internal Geneva wheel).

ω = constant angular velocity of driving crank, rad/sec = $\dfrac{p\pi}{30}$ rad/sec

α = angular position of driving crank at any time, degrees

β = angular displacement of driven member corresponding to crank angle α

$$\cos \beta = \dfrac{m + \cos \alpha}{\sqrt{1 + m^2 + 2m \cos \alpha}}$$

$$\text{Angular velocity of driven member} = \dfrac{d\beta}{dt} = \omega\left(\dfrac{1 + m \cos \alpha}{1 + m^2 + 2m \cos \alpha}\right)$$

$$\text{Angular acceleration of driven member} = \dfrac{d^2\beta}{dt^2} = \omega^2\left[\dfrac{m \sin \alpha(1 - m^2)}{(1 + m^2 + 2m \cos \alpha)^2}\right]$$

Maximum angular velocity occurs at $\alpha = 0°$ and equals $\dfrac{\omega}{1 + m}$ rad/sec

Maximum angular acceleration occurs when roller enters slot and equals

$$\dfrac{\omega^2}{\sqrt{m^2 - 1}} \text{ rad/sec}^2$$

MECHANISMS, LINKAGE GEOMETRY, AND CALCULATIONS **11.5**

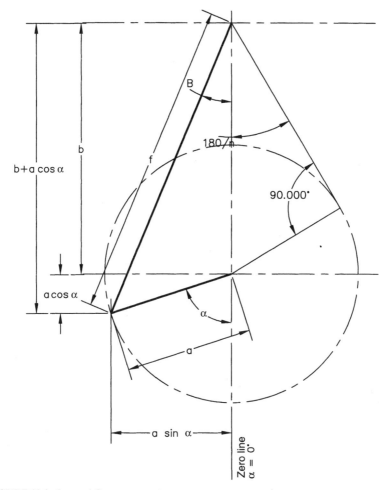

FIGURE 11.4 Internal Geneva geometry.

11.3 STANDARD MECHANISMS

- Figure 11.5 shows the scotch yoke mechanism for generating sine and cosine functions.
- Figure 11.6 shows the tangent and cotangent functions.
- Figure 11.7 shows the formulas for the roller-detent mechanism.
- Figure 11.8 shows the formulas for the plunger-detent mechanism.
- Figure 11.9 shows the slider-crank mechanism.

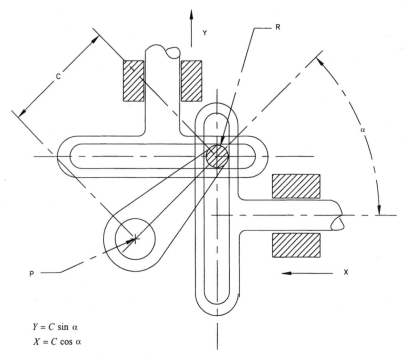

$Y = C \sin \alpha$
$X = C \cos \alpha$

FIGURE 11.5 Scotch yoke mechanism for sine and cosine functions.

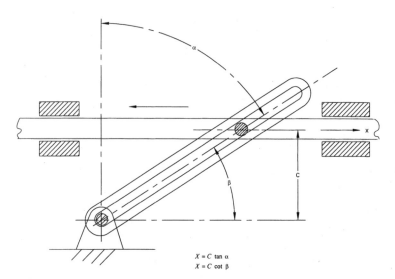

$X = C \tan \alpha$
$X = C \cot \beta$

FIGURE 11.6 Tangent-cotangent mechanism.

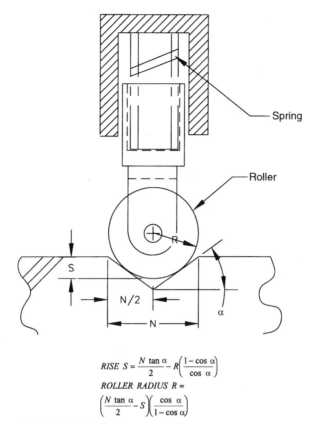

$$\text{RISE } S = \frac{N \tan \alpha}{2} - R\left(\frac{1-\cos \alpha}{\cos \alpha}\right)$$

ROLLER RADIUS $R =$
$$\left(\frac{N \tan \alpha}{2} - S\right)\left(\frac{\cos \alpha}{1-\cos \alpha}\right)$$

FIGURE 11.7 Roller-detent mechanism.

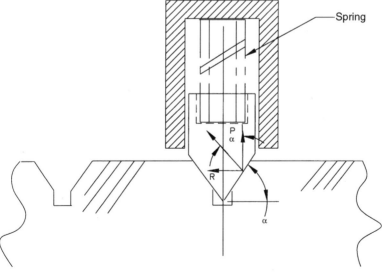

FIGURE 11.8 Plunger-detent mechanism. Holding power $R = P \tan \alpha$. For friction coefficient F at contact surface, $R = P (\tan \alpha + F)$.

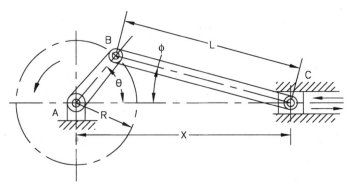

Displacement of slider:

$$X = L \cos \phi + R \cos \theta \qquad \cos \phi = \sqrt{\left[1 - \left(\frac{R}{L}\right)^2 \sin^2 \theta\right]}$$

Angular velocity of connecting rod:

$$\phi' = \omega \left[\frac{(R/L) \cos \phi}{[1 - (R/L)^2 \sin^2 \theta]^{1/2}}\right]$$

Linear velocity of piston:

$$X' = -\omega \left(\frac{1 + \phi'}{\omega}\right) \left(\frac{R}{L}\right) \sin \theta L$$

Angular acceleration of connecting rod:

$$\phi'' = \frac{\omega^2 (R/L) \sin \theta \,[(R/L^2) - 1]}{[1 - (R/L^2) \sin^2 \theta]^{3/2}}$$

Slider acceleration:

$$X'' = -\omega^2 \left(\frac{R}{L}\right) \left[\cos \theta + \frac{\phi''}{\omega^2} \sin \theta + \frac{\phi'}{\omega} \cos \theta\right] L$$

where L = length of connecting rod
R = Radius of crank
X = distance from center of crankshaft A to wrist pin C
X' = slider velocity (linear velocity of point C)
X'' = Slider acceleration
θ = crank angle measured from dead center when slider is fully extended
ϕ = angular position of connecting rod; $\phi = 0$ when $\theta = 0$
ϕ' = connecting rod angular velocity = $d\phi/dt$
ϕ'' = connecting rod angular acceleration = $d^2\phi/dt^2$
ω = constant angular velocity of the crank

FIGURE 11.9 Slider-crank mechanism.

11.4 CLAMPING MECHANISMS AND CALCULATION PROCEDURES

Clamping mechanisms are an integral part of nearly all tooling fixtures. Countless numbers of clamping designs may be used by the tooling fixture designer and toolmaker, but only the basic types are described in this section. With these basic clamp types, it is possible to design a vast number of different tools. Both manual and pneumatic/hydraulic clamping mechanisms are shown, together with the equations used to calculate each basic type. The forces generated by the pneumatic and hydraulic mechanisms may be calculated initially by using pneumatic and hydraulic formulas or equations.

The basic clamping mechanisms used by many tooling fixture designers are outlined in Fig. 11.10, types 1 through 12. These basic clamping mechanisms also may be used for other mechanical design applications.

Eccentric Clamp, Round (Fig. 11.10, Type 12). The eccentric clamp, such as that shown in Fig. 11.10, type 12, is a fast-action clamp compared with threaded clamps, but threaded clamps have higher clamping forces. The eccentric clamp usually develops clamping forces that are 10 to 15 times higher than the force applied to the handle.

The ratio of the handle length to the eccentric radius normally does not exceed 5 to 6, while for a swinging clamp or strap clamp (threaded clamps), the ratio of the handle length to the thread pitch diameter is 12 to 15. The round eccentrics are relatively cheap and have a wide range of applications in tooling.

The angle α in Fig. 11.10, type 12, is the rising angle of the round eccentric clamp. Because this angle changes with rotation of the eccentric, the clamping force is not proportional at all handle rotation angles. The clamping stroke of the round eccentric at 90° of its handle rotation equals the roller eccentricity e. The machining allowance for the clamped part or blank x must be less than the eccentricity e. To provide secure clamping, eccentricity $e \geq x$ to $1.5x$ is suggested.

The round eccentric clamp is supposed to have a self-holding characteristic to prevent loosening in operation. This property is gained by choosing the correct ratio of the roller diameter D to the eccentricity e. The holding ability depends on the coefficient of static friction. In design practice, the coefficient of friction f would normally be 0.1 to 0.15, and the self-holding quality is maintained when f exceeds $\tan \alpha$.

The equation for determining the clamping force P is

$$P = Ql \frac{l}{[\tan(\alpha + \phi_1) + \tan \phi_2]r}$$

Then the necessary handle torque ($M = Pl$) is

$$M = P[\tan(\alpha + \phi_1) + \tan \phi_2]r$$

where r = distance from pivot point to contact point of the eccentric and the machined part surface, in or mm

Type	Geometry	Equation
1		$Q = P\left(\dfrac{l + rf_0}{l_1 - rf_0}\right)$ $P = \dfrac{Q}{(l + rf_0)/(l_1 - rf_0)}$
2		$Q = P\left(\dfrac{l + hf + rf_0}{l_1 - h_1 f_1 - rf_0}\right)$ when $l_1 \geq l$ and $P \geq Q$

FIGURE 11.10 Clamping mechanisms.

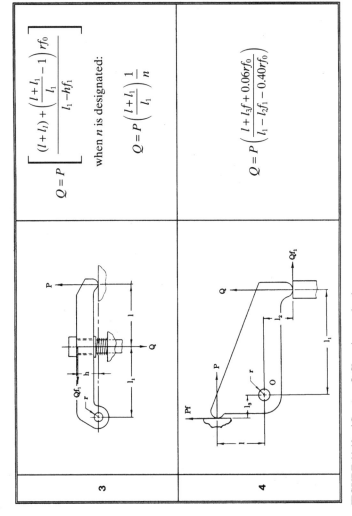

FIGURE 11.10 (*Continued*) Clamping mechanisms.

5	$Q = \left[P\left(\dfrac{l+l_1}{l_1}\right) + q \right] \dfrac{l_2}{l_3} \cdot \dfrac{1}{n}$ $Q = Q_0 \left(\dfrac{l_2}{l_3}\right) \cdot \dfrac{1}{n}$ and $Q_0 = P\left(\dfrac{l+l_1}{l_1}\right) + q$ q = spring resistance, lbf or N
6	$Q = 2P \tan(\alpha + \beta) \tan \alpha_1$ Where $\beta = \arcsin f_0$

FIGURE 11.10 (*Continued*) Clamping mechanisms.

11.12

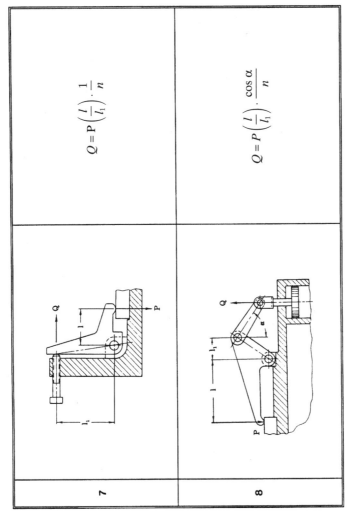

FIGURE 11.10 (*Continued*) Clamping mechanisms.

9	$Q = P\left(\dfrac{l}{l_1}\right) \cdot \dfrac{1}{n}$
10	$Q = P\left(\dfrac{\sin \alpha_1 l + \cos \alpha_1 h}{l_1}\right) \cdot \dfrac{1}{n}$

FIGURE 11.10 (*Continued*) Clamping mechanisms.

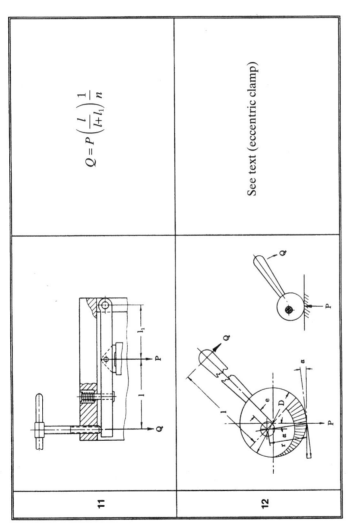

FIGURE 11.10 (*Continued*) Clamping mechanisms.

Note: f_0 = coefficient of friction (axles and pivot pins) = 0.1 to 0.15; f = coefficient of friction of clamped surface = tan ϕ; ϕ = arctan f; n = efficiency coefficient, 0.98 to 0.84, determined by frictional losses in pivots and bearings, 0.98 for the best bearings through 0.84 for no bearings (in order to avoid the use of complex, lengthy equations, the value of n can be taken as a mean between the limits shown); q = spring resistance or force, lbf or N.

α = rotation angle of the eccentric at clamping (reference only)
tan ϕ_1 = friction coefficient at the clamping point
tan ϕ_2 = friction coefficient in the pivot axle
  l = handle length, in or mm
  Q = force applied to handle, lbf or N
  D = diameter of eccentric blank or disc, in or mm
  P = clamping force, lbf or N

NOTE. tan $(\alpha + \phi_1) \approx 0.2$ and tan $\phi_2 \approx 0.05$ in actual practice.
See Fig. 11.11 for listed clamping forces for the eccentric clamp shown in Fig. 11.10, type 12.

The Cam Lock. Another clamping device that may be used instead of the eccentric clamp is the standard cam lock. In this type of clamping device, the clamping action is more uniform than in the round eccentric, although it is more difficult to manufacture. A true camming action is produced with this type of clamping device. The method for producing the cam geometry is shown in Fig. 11.12. The layout shown is for a cam surface generated in 90° of rotation of the device, which is the general application. Note that the cam angle should not exceed 9° in order for the clamp to function properly and be self-holding. The cam wear surface should be hardened to approximately Rockwell C30 to C50, or according to the application and the hardness of the materials which are being clamped. The cam geometry may be developed using CAD, and the program for machining the cam lock may be loaded into the CNC of a wire EDM machine.

FIGURE 11.11 Torque values for listed clamping forces—eccentric clamps (type 12, Fig. 11.10).

	Clamping force P, N						
D	490	735	980	1225	1470	1715	1960
40 mm (1.58 in)	2.65	3.97	5.40	6.67	8.00	9.37	10.64
50 mm (1.98 in)	3.34	5.00	6.67	8.39	10.01	11.77	13.68
60 mm (2.36 in)	4.02	6.03	8.00	10.01	11.97	14.03	16.48
70 mm (2.76 in)	4.71	7.06	9.42	11.77	14.08	16.48	18.79

Note: Tabulated values are torques, N · m.
To convert clamping forces in newtons to pounds force, multiply table values by 0.2248 · (i.e., 1960 N = 1960 × 0.2248 = 441 lbf).
To convert tabulated torques in newton-meters to pound-feet, multiply values by 0.7376 (i.e., 18.79 N · m = 18.79 × 0.7376 = 13.9 lb · ft).

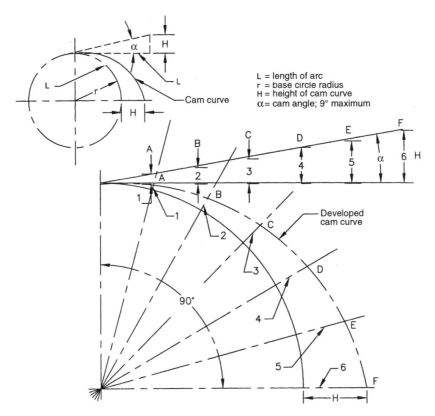

FIGURE 11.12 Cam lock geometry.

11.5 LINKAGES—SIMPLE AND COMPLEX

Linkages are an important element of machine design and are therefore detailed in this section, together with their mathematical solutions. Some of the more commonly used linkages are shown in Figs. 11.13 through 11.17. By applying these linkages to applications containing the simple machines, a wide assortment of workable mechanisms may be produced.

Toggle-Joint Linkages. Figure 11.13 shows the well-known and often-used toggle mechanism. The mathematical relationships are shown in the figure. The famous Luger pistol action is based on the toggle-joint mechanism.

The Four-Bar Linkage. Figure 11.14 shows the very important four-bar linkage, which is used in countless mechanisms. The linkage looks simple, but it was not until the 1950s that a mathematician was able to find the mathematical relationship between this linkage and all its parts. The equational relationship of the four-bar linkage is known as the *Freudenstein relationship* and is shown in the figure. The geometry of the linkage

11.18 CHAPTER ELEVEN

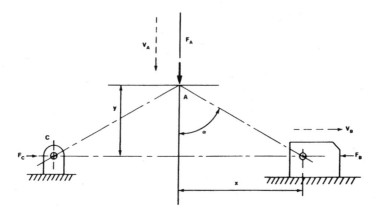

$$M_a = \frac{F_B}{F_A} = \frac{1}{2} \cdot \frac{x}{y} = \frac{1}{2}\tan\alpha = \frac{V_A}{V_B}.$$

As angle α approaches 90°, the links come into toggle, and the mechanical advantage and velocity ratio both approach infinity.

M_a = Mechanical advantage (ratio)
F_B = Force at point B
F_A = Force at point A
V_A = Velocity at point A
V_B = Velocity at point B
X = Horizontal displacement
Y = Vertical displacement

FIGURE 11.13 Toggle joint mechanism.

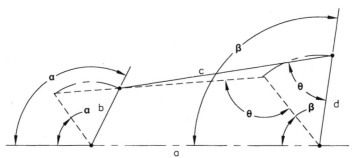

$$L_1 \cos\alpha - L_2 \cos\beta + L_3 = \cos(\alpha - \beta)$$

Where:

$$L_1 = \left(\frac{a}{d}\right) \qquad L_2 = \frac{a}{b} \qquad L_3 = \frac{b^2 - c^2 + d^2 + a^2}{2bd}$$

or

$$\frac{a}{d}\cos\alpha - \frac{a}{b}\cos\beta + \frac{b^2 - c^2 + d^2 + a^2}{2bd} = \cos(\alpha - \beta)$$

FIGURE 11.14 Four-bar mechanism. *a, b, c,* and *d* are the links. Angle α is the link *b* angle, and angle β is the follower link angle for link *d*. When links *a, b,* and *d* are known, link *c* can be calculated as shown. The transmission angle θ can also be calculated using the equations.

MECHANISMS, LINKAGE GEOMETRY, AND CALCULATIONS   11.19

may be ascertained with the use of trigonometry, but the velocity ratios and the actions are extremely complex and can be solved only using advanced mathematics.

The use of high-speed photography on a four-bar mechanism makes its analysis possible without recourse to advanced mathematical methods, provided that the mechanism can be photographed.

Simple Linkages.   In Fig. 11.15, the torque applied at point T is known, and we wish to find the force along link F. We proceed as follows: First, find the effective value of force F_1, which is:

$$F_1 \times R = T$$

$$F_1 = \frac{T}{R}$$

Then

$$\sin \phi = \frac{F_1}{F}$$

$$F = \frac{F_1}{\sin \phi} \quad \text{or} \quad \frac{T/R}{\sin \phi}$$

NOTE. $T/R = F_1$ = torque at T divided by radius R.

In Fig. 11.16, the force F acting at an angle θ is known, and we wish to find the torque at point T. First, we determine angle α from $\alpha = 90° - \theta$ and then proceed to find the vector component force F_1, which is

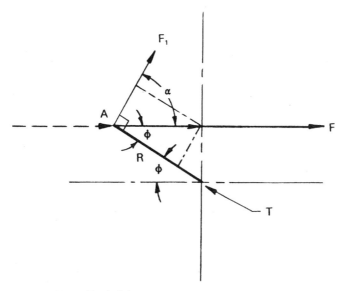

FIGURE 11.15   Simple linkage.

$$\cos \alpha = \frac{F_1}{F} \quad \text{and} \quad F_1 = F \cos \alpha$$

The torque at point T is $F \cos \alpha R$, which is $F_1 R$. (Note that F_1 is at 90° to R.)

Crank Linkage. In Fig. 11.17, a downward force F will produce a vector force F_1 in link AB. The instantaneous force at 90° to the radius arm R, which is P_n, will be

$$F_1 \quad \text{or} \quad P = \frac{F}{\cos \alpha}$$

and

$$P_n = F_1 \quad \text{or} \quad P \cos \lambda \quad \text{or} \quad P_n = \frac{F}{\cos \phi} \sin (\phi - \theta)$$

The resulting torque at T will be $T = P_n R$, where R is the arm BT.

The preceding case is typical of a piston acting through a connecting rod to a crankshaft. This particular linkage is used many times in machine design, and the applications are countless.

The preceding linkage solutions have their roots in engineering mechanics, further practical study of which may be made using the McGraw-Hill *Electromechanical Design Handbook, Third Edition* (2000), also written by the author.

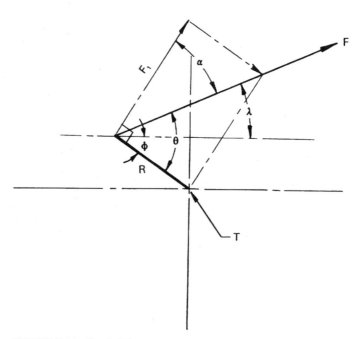

FIGURE 11.16 Simple linkage.

MECHANISMS, LINKAGE GEOMETRY, AND CALCULATIONS 11.21

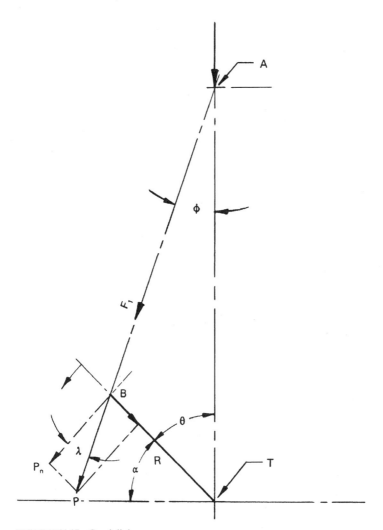

FIGURE 11.17 Crank linkage.

Four-Bar Linkage Solutions Using a Hand-Held Calculator. Figure 11.14 illustrates the standard Freudenstein equation which is the basis for deriving the very important four-bar linkage used in many engineering mechanical applications. Practical solutions using the equation were formerly limited because of the complex mathematics involved. Such computations have become readily possible, however, with the advent of the latest generation of hand-held programmable calculators, such as the Texas Instruments TI-85 and the Hewlett Packard HP-48G. Both of these new-generation calculators operate like small computers, and both have enormous capabilities in solving general and very difficult engineering mathematics problems.

Refer to Fig. 11.14 for the geometry of the four-bar linkage. The short form of the general four-bar-linkage equation is:

$$L_1 \cos \alpha - L_2 \cos \beta + L_3 = \cos(\alpha - \beta)$$

where $L_1 = a/d$
$L_2 = a/b$
$L_3 = b^2 - c^2 + d^2 + a^2/2bd$

The correct working form for the equation is:

$$\frac{a}{d}\cos\alpha - \frac{a}{b}\cos\beta + \frac{b^2 - c^2 + d^2 + a^2}{2bd} = \cos(\alpha - \beta)$$

Transposing the equation to solve for c, we obtain:

$$c = \left[\left(\left\{[-\cos(\alpha-\beta)] + \left(\frac{a}{d}\right)\cos\alpha - \left(\frac{a}{b}\right)\cos\beta\right\}2bd\right) + d^2 + b^2 + a^2\right]^{0.5}$$

This equation must be entered into the calculator as shown, except that the brackets and braces must be replaced by parentheses in the calculator. If the equation is not correctly separated with parentheses according to the proper algebraic order of operations, the calculator will give an error message. Thus, on the TI-85 the equation must appear as shown here:

$$c = ((((-\cos(A-B)) + (R/T)\cos A - (R/S)\cos B)(2ST)) + s^2 + T^2 + R^2)^{0.5}$$

NOTE. $A = \alpha, B = \beta, R = a, T = d,$ and $S = b$. (The TI-85 cannot show $\alpha, \beta, a, b,$ and d.) When we know angle α, angle β may be solved by:

$$\beta = \cos^{-1}\frac{h^2 + a^2 - b^2}{2ha} + \cos^{-1}\frac{h^2 + d^2 - c^2}{2hd}$$

where $h^2 = (a^2 + b^2 + 2ab\cos\alpha)$
$h = (a^2 + b^2 + 2ab\cos\alpha)^{0.5}$

The transmission angle θ is therefore:

$$\theta = \cos^{-1}\frac{c^2 + d^2 - a^2 - b^2 - 2ab\cos\alpha}{2cd}$$

In the figure, the driver link is b and the driven link is d. When driver link b moves through a different angle α, we may compute the final follower angle β and the transmission angle θ.

The equation for the follower angles β, shown previously, must be entered into the calculators as shown here (note that $\cos^{-1} = \arccos$):

$$\beta = (\cos^{-1}((H^2 + R^2 - S^2)/(2HR))) + (\cos^{-1}((H^2 + T^2 - K^2)/(2HT)))$$

and the transmission angles θ must be entered as shown here:

MECHANISMS, LINKAGE GEOMETRY, AND CALCULATIONS   11.23

$$\theta = \cos^{-1}((K^2 + T^2 - R^2 - S^2 - 2RS \cos A)/(2KT))$$

As before, the capital letters must be substituted for actual equation letters as codes.

NOTE. In the preceding calculator entry form equations, the calculator exponent symbols (^) have been omitted for clarity; for example,

$$\cos^{-1}((K^{\wedge}2 + T^{\wedge}2 - R^{\wedge}2 \ldots))$$

It is therefore of great importance to learn the proper entry and bracketing form for equations used on the modern calculators, as illustrated in the preceding explanations and in Sec. 1.4.

Figure 11.18 shows a printout from the MathCad PC program, which presents a complete mathematical solution of a four-bar linkage. As a second proof of the problem shown in Fig. 11.18, the linkage was drawn to scale using AutoCad LT in Fig. 11.19. As can be seen from these two figures, the basic Freudenstein equations are mathematically exact.

Freudenstein's Equation for 4-Bar Linkages

$\alpha := \deg 95 \quad \beta := \deg 98 \quad b := 1.875 \quad d := 2.500 \quad a := 12.625$

$$\left[\left[-\cos(\alpha - \beta) + \cos(\alpha)\cdot\left(\frac{a}{d}\right) - \cos(\beta)\cdot\left(\frac{a}{b}\right)\right]\cdot 2\cdot b\cdot d + (d^2 + b^2 + a^2)\right]^{0.5} = 12.823928$$

The above equation for solving the "c" link of a 4-bar equation was transposed from the Freudenstein relational equation shown below. The other important relational equations for angles and sides follow the Freudenstein equation.

From the above, $c := 12.823928$ = the unknown link length (c).

Freudenstein's Equation: Standard form.

$$\left(\cos(\alpha)\cdot\frac{a}{d} - \cos(\beta)\cdot\frac{a}{b}\right) + \frac{(b^2 - c^2 + d^2 + a^2)}{2\cdot b\cdot d} = \cos(\alpha - \beta)$$

FIGURE 11.18   Four-bar linkage solved by MathCad.

Relational Equations: $h := 12.600792 \quad \alpha := 95 \text{ deg} \quad \beta := 98$

When we know angle α, angle β may be solved by the following when:

$$h^2 = 158.779959 \quad \text{and} \quad (a^2 + b^2 + 2 \cdot a \cdot b \cdot \cos(\alpha))^{0.5} = 12.600792 = h$$

$$\beta = \text{acos}\left(\frac{h^2 + a^2 - b^2}{2 \cdot h \cdot a}\right) + \text{acos}\left(\frac{h^2 + d^2 - c^2}{2 \cdot h \cdot d}\right) = 98.000018 \bullet \text{deg}$$

This checks with β above within 0.000018 degrees, or 0.06 seconds.

The transmission angle θ is therefore:

$\cos(\alpha) = -0.087156$

$$\theta = \text{acos}\left(\frac{c^2 + d^2 - a^2 - b^2 - 2 \cdot a \cdot b \cdot \cos(\alpha)}{2 \cdot c \cdot d}\right) = 79.283374 \bullet \text{deg}$$

FIGURE 11.18 Four-bar linkage solved by MathCad.

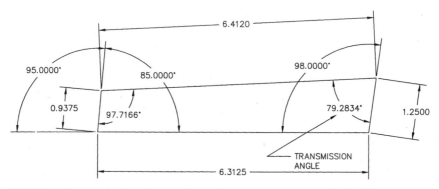

FIGURE 11.19 A scaled AutoCad drawing confirming calculations shown in Fig. 11.18.

CHAPTER 12
CLASSES OF FITS FOR MACHINED PARTS— CALCULATIONS

12.1 CALCULATING BASIC FIT CLASSES (PRACTICAL METHOD)

The following examples of calculations for determining the sizes of cylindrical parts fit into holes were accepted as an industry standard before the newer U.S. customary and ISO fit standards were established. This older method is still valid when part tolerance specifications do not require the use of the newer standard fit classes. Refer to Fig. 12.1 for the tolerances and allowances shown in the following calculations.

From Fig. 12.1a, upper and lower fit limits are selected for a class A hole and a class Z shaft of 1.250-in nominal diameter.

For the class A hole:

$$1.250 \text{ in} - 0.00025 \text{ in} = \text{high limit} = 1.25025 \text{ in}$$

$$1.250 \text{ in} - 0.00150 \text{ in} = \text{low limit} = 1.24975 \text{ in}$$

The hole dimension will then be 1.24975- to 1.25025-in diameter (See Fig. 12.2).
For a class Z fit of the shaft:

$$1.250 \text{ in} - 0.00075 \text{ in} = \text{high limit} = 1.24925 \text{ in}$$

$$1.250 \text{ in} - 0.00150 \text{ in} = \text{low limit} = 1.24850 \text{ in}$$

The shaft dimension will then be 1.24925- to 1.24850-in diameter (See Fig. 12.2). The minimum and maximum clearances will then be:

1.24975 in = min. hole dia.	1.25025 in = max. hole dia.
− 1.24925 in = max. shaft dia.	− 1.24850 in = min. shaft dia.
0.00050 in minimum clearance	0.00175 in maximum clearance

Allowances for Fits—Bearings and Other Cylindrical Machined Parts

Class	Nominal diameter	Up to 0.500 in	0.5625–1 in	1.0625–2 in	2.0625–3 in	3.0625–4 in	4.0625–5 in
		Tolerances in standard holes*					
A	High limit	+0.00025	+0.0005	+0.00075	+0.0010	+0.0010	+0.0010
	Low limit	−0.00025	−0.00025	−0.00025	−0.0005	−0.0005	−0.0005
	Tolerance	0.0005	0.00075	0.0010	0.0015	0.0015	0.0015
B	High limit	+0.0005	+0.00075	+0.0010	+0.00125	+0.0015	+0.00175
	Low limit	−0.0005	−0.0005	−0.0005	−0.00075	−0.00075	−0.00075
	Tolerance	0.0010	0.00125	0.0015	0.0020	0.00225	0.0025
		Allowances for forced fits					
F	High limit	+0.0010	+0.0020	+0.0040	+0.0060	+0.0080	+0.0100
	Low limit	+0.0005	+0.0015	+0.0030	+0.0045	+0.0060	+0.0080
	Tolerance	0.0005	0.0005	0.0010	0.0015	0.0020	0.0020
		Allowances for driving fits					
D	High limit	+0.0005	+0.0010	+0.0015	+0.0025	+0.0030	+0.0035
	Low limit	+0.00025	+0.00075	+0.0010	+0.0015	+0.0020	+0.0025
	Tolerance	0.00025	0.00025	0.0005	0.0010	0.0010	0.0010
		Allowances for push fits					
P	High limit	−0.00025	−0.00025	−0.00025	−0.0005	−0.0005	−0.0005
	Low limit	−0.00075	−0.00075	−0.00075	−0.0010	−0.0010	−0.0010
	Tolerance	0.0005	0.0005	0.0005	0.0005	0.0005	0.0005
		Allowances for running fits†					
X	High limit	−0.0010	−0.00125	−0.00175	−0.0020	−0.0025	−0.0030
	Low limit	−0.0020	−0.00275	−0.0035	−0.00425	−0.0050	−0.00575
	Tolerance	0.0010	0.0015	0.00175	0.00225	0.0025	0.00275

FIGURE 12.1 Allowances for fits (common practice).

Allowances for running fits[†] (Continued)

Class	Nominal diameter	Up to 0.500 in	0.5625–1 in	1.0625–2 in	2.0625–3 in	3.0625–4 in	4.0625–5 in
Y	High limit	−0.00075	−0.0010	−0.00125	−0.0015	−0.0020	−0.00225
	Low limit	−0.00125	−0.0020	−0.0025	−0.0030	−0.0035	−0.0040
	Tolerance	0.0005	0.0010	0.00125	0.0015	0.0015	0.00175
Z	High limit	−0.0005	−0.00075	−0.00075	−0.0010	−0.0010	−0.00125
	Low limit	−0.00075	−0.00125	−0.0015	−0.0020	−0.00225	−0.0025
	Tolerance	0.00025	0.0005	0.00075	0.0010	0.00125	0.00125

(a)

Class	High limit	Low limit
A	$+ D^{0.5} \times 0.0006$	$- D^{0.5} \times 0.0003$
B	$+ D^{0.5} \times 0.0008$	$- D^{0.5} \times 0.0004$
P	$+ D^{0.5} \times 0.0002$	$- D^{0.5} \times 0.0006$
X	$+ D^{0.5} \times 0.00125$	$- D^{0.5} \times 0.0025$
Y	$+ D^{0.5} \times 0.001$	$- D^{0.5} \times 0.0018$
Z	$+ D^{0.5} \times 0.0005$	$- D^{0.5} \times 0.001$

Note: D = basic diameter of part, in.

(b)

FIGURE 12.1 (*Continued*) Allowances for fits (common practice).

*Tolerance is provided for holes which ordinary standard reamers can produce, in two grades, class A and B, the selection of which is a question for the user's decision and dependent upon the quality of the work required. Some prefer to use class A as working limits and class B as inspection limits.

[†] Running fits, which are the most commonly required, are divided into three grades: class X, for engine and other work where easy fits are desired; class Y, for high speeds and good average machine work; and class Z, for fine tooling work.

12.4 CHAPTER TWELVE

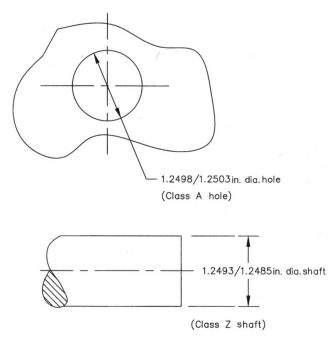

FIGURE 12.2 Class A hole to class Z shaft fit dimensions.

The hole and shaft dimensions may be rounded to 4 decimal places for a more practical application.

NOTE. In using Fig. 12.1a and b, class A and B entries are for the holes, and all the other classes are used for the shaft or other cylindrical parts. You may also use Fig. 12.1b to calculate the upper and lower limits for holes and cylindrical parts, using the equations shown in the figure.

Problem. Using Fig. 12.1a, find the hole- and bearing-diameter dimensions for a bearing of 1.7500 in OD to be a class D driving or arbor press fit in a class A bored hole.

Solution. From Fig. 12.1a, the class A hole for a 1.750-in-diameter bearing is:

$$1.7500 \text{ in} + 0.00075 \text{ in} = \text{high limit} = 1.75075 \text{ in}$$

$$1.7500 \text{ in} - 0.00025 \text{ in} = \text{low limit} = 1.74975 \text{ in}$$

The hole dimension is therefore 1.74975- to 1.75075-in diameter.
The bearing diameter for a class D driving or press fit is:

$$1.7500 \text{ in} + 0.0015 \text{ in} = \text{high limit} = 1.7515 \text{ in}$$

$$1.7500 \text{ in} + 0.0010 \text{ in} = \text{low limit} = 1.7510 \text{ in}$$

The bearing OD dimension is therefore 1.7515- to 1.7510-in diameter.

CLASSES OF FITS FOR MACHINED PARTS—CALCULATIONS   **12.5**

The minimum and maximum interferences are then:

    1.75100 in min. bearing dia.   1.75150 in max. bearing dia.

    − 1.75075 in max. bore dia.   − 1.74975 in min. bore dia.

    0.00025 in min. interference   0.00175 in max. interference

Rounded to 4 decimal places:

    0.0003 in minimum interference   0.0018 in maximum interference

For the new U.S. customary and ISO fit classes and their calculations, see Sec. 12.2.

12.2 U.S. CUSTOMARY AND METRIC (ISO) FIT CLASSES AND CALCULATIONS

Limits and fits of shafts and holes are important design and manufacturing considerations. Fits should be carefully selected according to function. The fits outlined in this section are all on a unilateral hole basis. Table 12.1 describes the various U.S. customary fit designations. Classes RC9, LC10, and LC11 are described in the ANSI standards but are not included here. Table 12.1 is valid for sizes up to approximately 20 in diameter and is in accordance with American, British, and Canadian recommendations.

The coefficients C listed in Table 12.2 are to be used with the equation $L = CD^{1/3}$, where L is the limit in thousandths of an inch corresponding to the coefficients C and the basic size D in inches. The resulting calculated values of L are then summed algebraically to the basic shaft size to obtain the four limiting dimensions for the shaft and hole. The limits obtained by the preceding equation and Table 12.2 are very close approximations to the standards, and are applicable in all cases except where exact conformance to the standards is required by specifications.

EXAMPLE. A *precision running fit* is required for a nominal 1.5000-in-diameter shaft (designated as an RC3 fit per Table 12.2).

    Lower Limit for the Hole   *Upper Limit for the Hole*

$$L_1 = \frac{CD^{1/3}}{1000} \qquad L_2 = \frac{CD^{1/3}}{1000}$$

$$L_1 = \frac{0\,(1.5)^{1/3}}{1000} \qquad L_2 = \frac{0.907\,(1.5)^{1/3}}{1000}$$

$$L_1 = 0 \qquad L_2 = \frac{1.03825}{1000}$$

$$d_L = 0 + 1.5000 \qquad d_U = 0.001038 + 1.5000$$

$$d_L = 1.50000 \qquad d_U = 1.50104$$

TABLE 12.1 U.S. Customary Fit Class Designations

Designation	Name and application
RC1	*Close sliding fits* are intended for accurate location of parts which must be assembled without perceptible play.
RC2	*Sliding fits* are intended for accurate location, but with greater maximum clearance than the RC1 fit.
RC3	*Precision running fits* are the loosest fits that can be expected to run freely. They are intended for precision work at slow speeds and light pressures, but are not suited for temperature differences.
RC4	*Close-running fits* are intended for running fits on accurate machinery with moderate speeds and pressures. They exhibit minimum play.
RC5	*Medium-running fits* are intended for higher running speeds or heavy journal pressures, or both.
RC6	*Medium-running fits* are for use where more play than RC5 is required.
RC7	*Free-running fits* are for use where accuracy is not essential or where large temperature variations may occur, or both.
RC8	*Loose-running fits* are intended where wide commercial tolerances may be necessary, together with an allowance on the hole.
LC1 to LC9	*Locational-clearance fits* are required for parts which are normally stationary, but which can be freely assembled and disassembled. Snug fits are for accuracy of location. Medium fits are for parts such as ball, race, and housing. The looser fastener fits are used where freedom of assembly is important.
LT1 to LT6	*Locational-transitional fits* are a compromise between clearance and interference fits where accuracy of location is important, but either a small amount of clearance or interference is permitted.
LN1 to LN3	*Locational-interference fits* are for accuracy of location and for parts requiring rigidity and alignment, with no special requirement for bore pressure. Not intended for parts that must transmit frictional loads to one another.
FN1	*Light-drive fits* require light assembly pressures and produce permanent assemblies. Suitable for thin sections or long fits or in cast-iron external members.
FN2	*Medium-drive fits* are for ordinary steel parts or shrink fits on light sections. They are the tightest fits that can be used with high-grade cast-iron external members.
FN3	*Heavy-drive fits* are for heavier steel parts or for shrink fits in medium sections.
FN4 and FN5	*Force fits* are suitable for parts which can be highly stressed or for shrink fits where the heavy pressing forces required are not practical.

CLASSES OF FITS FOR MACHINED PARTS—CALCULATIONS

TABLE 12.2 Coefficient C for Fit Equations

Class of fit	Hole limits Lower	Hole limits Upper	Shaft limits Lower	Shaft limits Upper
RC1	0	0.392	−0.588	−0.308
RC2	0	0.571	−0.700	−0.308
RC3	0	0.907	−1.542	−0.971
RC4	0	1.413	−1.879	−0.971
RC5	0	1.413	−2.840	−1.932
RC6	0	2.278	−3.345	−1.932
RC7	0	2.278	−4.631	−3.218
RC8	0	3.570	−7.531	−5.253
LC1	0	0.571	−0.392	0
LC2	0	0.907	−0.571	0
LC3	0	1.413	−0.907	0
LC4	0	3.570	−2.278	0
LC5	0	0.907	−0.879	−0.308
LC6	0	2.278	−2.384	−0.971
LC7	0	3.570	−4.211	−1.933
LC8	0	3.570	−5.496	−3.218
LC9	0	5.697	−8.823	−5.253
LT1	0	0.907	−0.281	0.290
LT2	0	1.413	−0.442	0.465
LT3*	0	0.907	0.083	0.654
LT4*	0	1.413	0.083	0.990
LT5	0	0.907	0.656	1.227
LT6	0	0.907	0.656	1.563
LN1	0	0.571	0.656	1.048
LN2	0	0.907	0.994	1.565
LN3	0	0.907	1.582	2.153
FN1	0	0.571	1.660	2.052
FN2	0	0.907	2.717	3.288
FN3†	0	0.907	3.739	4.310
FN4	0	0.907	5.440	6.011
FN5	0	1.413	7.701	8.608

Note: Above coefficients for use with equation $L = CD^{1/3}$.
* Not for sizes under 0.24 in.
† Not for sizes under 0.95 in.
Source: Shigley and Mischke, *Standard Handbook of Machine Design*, McGraw-Hill, 1986.

Lower Limit for the Shaft

$$L_3 = \frac{CD^{1/3}}{1000}$$

$$L_3 = \frac{(-1.542)(1.5)^{1/3}}{1000}$$

$$L_3 = \frac{-1.76513}{1000}$$

$D_L = 1.500 + (-0.00176513)$

$D_L = 1.49823$

Upper Limit for the Shaft

$$L_4 = \frac{CD^{1/3}}{1000}$$

$$L_4 = \frac{(-0.971)(1.5)^{1/3}}{1000}$$

$$L_4 = \frac{-1.11150}{1000}$$

$D_U = 1.500 + (-0.0011115)$

$D_U = 1.49889$

Therefore, the hole and shaft limits are as follows:

$$\text{Hole size} = \frac{1.50000}{1.50104} \text{ dia.}$$

$$\text{Shaft size} = \frac{1.49889}{1.49823} \text{ dia.}$$

NOTE. Another often-used procedure for fit classes for shafts and holes is given in Fig. 12.1. Figure 12.1a shows tolerances in fits and Fig. 12.1b gives the equations for calculating allowances for the different classes of fits shown there. The procedures shown in Fig. 12.1 have often been used in industrial applications for bearing fits and fits of other cylindrical machined parts.

Table 12.3 shows the metric preferred fits for cylindrical parts in holes. The procedures for calculating the limits of fit for the metric standards are shown in the ANSI standards. The appropriate standard is ANSI B4.2—1978 (R1984). An alter-

TABLE 12.3 SI (Metric) Standard Fit Class Designations

Type	Hole basis	Shaft basis	Name and application
Clearance	H11/c11	C11/h11	*Loose-running fits* are for wide commercial tolerances or allowances on external parts.
	H9/d9	D9/h9	*Free-running fits* are not for use where accuracy is essential, but are good for large temperature variations, high running speeds, or heavy journal pressures.
	H8/f7	F8/h7	*Close-running fits* are for running on accurate machines and accurate location at moderate speeds and journal pressures.
	H7/g6	G7/h6	*Sliding fits* are not intended for running freely, but allow free movement and turning for accurate location.
	H7/h6	H7/h6	*Locational-clearance fits* provide snug fits for locating stationary parts, but can be freely assembled and disassembled.
Transition	H7/k6	K7/h6	*Locational-transition fits* are for accurate location, a compromise between clearance and interference.
	H7/n6	N7/h6	*Locational-transition fits* are for more accurate location where greater interference is permitted.
Interference	H7/p6	P7/h6	*Locational-interference fits* are for parts requiring rigidity and alignment with prime accuracy of location but with special bore pressures required.
	H7/s6	S7/h6	*Medium-drive fits* are for ordinary steel parts or shrink fits on light sections, the tightest fit usable with cast iron.
	H7/u6	U7/h6	*Force fits* are suitable for parts which can be highly stressed or for shrink fits where the heavy pressing forces required are not practical.

native to this procedure would be to correlate the type of fit between the metric standard fits shown in Table 12.3 with the U.S. customary fits shown in Table 12.1 and proceed to convert the metric measurements in millimeters to inches, and then calculate the limits of fit according to the method shown in this section for the U.S. customary system. The calculated answers would then be converted back to millimeters.

There should be no technical problem with this procedure except conflict with mandatory specifications, in which case you will need to concur with ANSI B4.2—1978(R1984) for the metric standard. The U.S. customary standard for preferred limits and fits is ANSI B4.1—1967(R1987).

The preceding procedures for limits and fits are mandatory practice for design engineers, tool design engineers, and toolmakers, in order for parts to function according to their intended design requirements. Assigning arbitrary or rule-of-thumb procedures to the fitting of cylindrical parts in holes is not good practice and can create many problems in the finished product.

12.3 CALCULATING PRESSURES, STRESSES, AND FORCES DUE TO INTERFERENCE FITS, FORCE FITS, AND SHRINK FITS

Interference- or Force-Fit Pressures and Stresses (Method 1). The stresses caused by interference fits may be calculated by considering the fitted parts as thick-walled cylinders, as shown in Fig. 12.3. The following equations are used to determine these stresses:

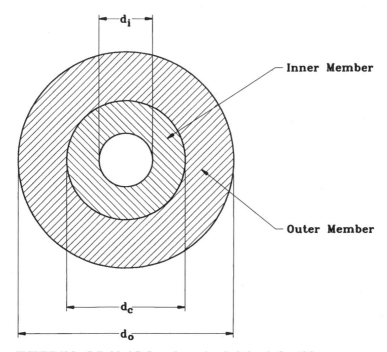

FIGURE 12.3 Cylindrical fit figure for use in calculations in Sec. 12.3.

$$P_c = \frac{\delta}{d_c \left[\dfrac{d_c^2 + d_i^2}{E_i(d_c^2 - d_i^2)} + \dfrac{d_o^2 + d_c^2}{E_o(d_o^2 - d_c^2)} - \dfrac{\mu_i}{E_i} + \dfrac{\mu_o}{E_o} \right]}$$

where P_c = pressure at the contact surface, psi
 δ = the total interference, in (diametral interference)
 d_i = inside diameter of the inner member, in
 d_c = diameter of the contact surface, in
 d_o = outside diameter of outer member, in
 μ_o = Poisson's ratio for outer member
 μ_i = Poisson's ratio for inner member
 E_o = modulus of elasticity of outer member, psi
 E_i = modulus of elasticity of inner member, psi

(See Table 12.4 for μ and E values.)

TABLE 12.4 Poisson's Ratio and Modulus of Elasticity Values

Material	Modulus of elasticity E, 10^6 psi	Poisson's ratio μ
Aluminum, various alloys	9.9–10.3	0.330–0.334
Aluminum, 6061-T6	10.2	0.35
Aluminum, 2024-T4	10.6	0.32
Beryllium copper	18	0.29
Brass, 70-30	15.9	0.331
Brass, cast	14.5	0.357
Bronze	14.9	0.14
Copper	15.6	0.355
Glass ceramic, machinable	9.7	0.29
Inconel	31	0.27–0.38
Iron, cast	13.5–21.0	0.221–0.299
Iron, ductile	23.8–25.2	0.26–0.31
Iron, grey cast	14.5	0.211
Iron, malleable	23.6	0.271
Lead	5.3	0.43
Magnesium alloy	6.3	0.281
Molybdenum	48	0.307
Monel metal	25	0.315
Nickel silver	18.5	0.322
Nickel steel	30	0.291
Phosphor bronze	13.8	0.359
Stainless steel, 18-8	27.6	0.305
Steel, cast	28.5	0.265
Steel, cold-rolled	29.5	0.287
Steel, all others	28.6–30.0	0.283–0.292
Titanium, 99.0 Ti	15–16	0.24
Titanium, Ti-8Al-1Mo-1V	18	0.32
Zinc, cast alloys	10.9–12.4	0.33
Zinc, wrought alloys	6.2–14	0.33

CLASSES OF FITS FOR MACHINED PARTS—CALCULATIONS   12.11

If the outer and inner members are of the same material, the equation reduces to:

$$P_c = \frac{\delta}{\dfrac{2d_c^3(d_o^2 - d_i^2)}{E(d_c^2 - d_i^2)(d_o^2 - d_c^2)}}$$

After P_c has been determined, then the actual tangential stresses at the various surfaces, in accordance with Lame's equation, for use in conjunction with the maximum shear theory of failure, may be determined by the following four equations:
On the surface at d_o:

$$S_{to} = \frac{2P_c d_c^2}{d_o^2 - d_c^2}$$

On the surface at d_c for the outer member:

$$S_{tco} = P_c\left(\frac{d_o^2 + d_c^2}{d_o^2 - d_c^2}\right)$$

On the surface at d_c for the inner member:

$$S_{tci} = -P_c\left(\frac{d_c^2 + d_i^2}{d_c^2 - d_i^2}\right)$$

On the surface at d_i:

$$S_{ti} = \frac{-2P_c d_c^2}{d_c^2 - d_i^2}$$

Interference-Fit Pressures and Stresses (Method 2).   The pressure for interference fit with reference to Fig. 12.3 is obtained from the following equations (symbol designations follow):

$$P = \frac{\delta}{b\dfrac{1}{E_i}\left(\dfrac{b^2 + a^2}{b^2 - a^2} - v_i\right) + \dfrac{1}{E_o}\left(\dfrac{c^2 + b^2}{c^2 - b^2} + v_o\right)} \qquad \text{(Eq. 12.1)}$$

If the inner cylinder is solid, then $a = 0$, and Eq. 12.1 becomes:

$$P = \frac{\delta}{b\dfrac{1}{E_i}(1 - v_i) + \dfrac{1}{E_o}\left(\dfrac{c^2 + b^2}{c^2 - b^2} + v_o\right)} \qquad \text{(Eq. 12.2)}$$

If the force-fit parts have identical moduli, Eq. 12.1 becomes:

$$P = \frac{E\delta}{b}\left[\frac{(c^2 - b^2)(b^2 - a^2)}{2b^2(c^2 - a^2)}\right] \qquad \text{(Eq. 12.3)}$$

If the inner cylinder is solid, Eq. 12.3 simplifies to become:

$$P = \frac{E\delta}{bc^2}(c^2 - b^2) \qquad \text{(Eq. 12.4)}$$

where P = pressure, psi
  δ = radial interference (total maximum interference divided by 2), in
  E = modulus of elasticity, Young's modulus (tension), 30×10^6 psi for most steels
  v = Poisson's ratio, 0.30 for most steels
  V_o = Poisson's ratio of outer member
  V_i = Poisson's ratio of inner member
  a, b, c = radii of the force-fit cylinders; $a = 0$ when the inner cylinder is solid (see Fig. 12.3)

NOTE. Equation 12.1 is used for two force-fit cylinders with different moduli; Eq. 12.2 is used for two force-fit cylinders with different moduli and the inner member is a solid cylinder; Eq. 12.3 is used in place of Eq. 12.1 if the moduli are identical; and Eq. 12.4 is used in place of Eq. 12.3 if the moduli are identical and the inner cylinder is solid, such as a shaft.

The maximum stresses occur at the contact surfaces. These are known as *biaxial stresses*, where t and r designate tangential and radial directions. Then, for the outer member, the stress is:

$$\sigma_{ot} = P\frac{c^2 + b^2}{c^2 - b^2} \qquad \text{while} \qquad \sigma_{or} = -P$$

For the inner member, the stresses at the contact surface are:

$$\sigma_{it} = -P\frac{b^2 + a^2}{b^2 - a^2} \qquad \text{while} \qquad \sigma_{ir} = -P$$

Use stress concentration factors of 1.5 to 2.0 for conditions such as a thick hub press-fit to a shaft. This will eliminate the possibility of a brittle fracture or fatigue failure in these instances.

(*Source:* Shigley and Mischke, *Standard Handbook of Machine Design*, McGraw-Hill, 1986.)

Forces and Torques for Force Fits. The maximum axial force F_a required to assemble a force fit varies directly as the thickness of the outer member, the length of the outer member, the difference in diameters of the force-fitted members, and the coefficient of friction. This force in pounds may be approximated with the following equation:

$$F_a = f\pi d L P_c$$

The torque that can be transmitted by an interference fit without slipping between the hub and shaft can be estimated by the following equation (parts must be clean and unlubricated):

$$T = \frac{fP_c \pi d^2 L}{2}$$

where F_a = axial load, lb
T = torque transmitted, lb · in
d = nominal shaft diameter, in
f = coefficient of static friction
L = length of external member, in
P_c = pressure at the contact surfaces, psi

Shrink-Fit Assemblies. Assembly of shrink-fit parts is facilitated by heating the outer member or hub until it has expanded by an amount at least as much as the diametral interference δ. The temperature change ΔT required to effect δ (diametral interference) on the outer member or hub may be determined by:

$$\Delta T = \frac{\delta}{\alpha d_i} \qquad \delta = \Delta T \alpha d_i \qquad d_i = \frac{\delta}{\alpha \Delta T}$$

where δ = diametral interference, in
α = coefficient of linear expansion per °F
ΔT = change in temperature on outer member above ambient or initial temperature, °F
d_i = initial diameter of the hole before expansion, in

An alternative to heating the hub or outer member is to cool the shaft or inner member by means of a coolant such as dry ice (solid CO_2) or liquid nitrogen.

INDEX

Algebra, 1.7–1.11
Algebraic procedures, 1.7
   bracketing in, 1.31–1.32
Angles, 1.32–1.33
   calculating, 4.15–4.20
   complex, finding, 4.54–4.63
   cutting, 4.58–4.60
   setting, 4.1–4.6

Boring calculations, 5.66–5.67
Boring coordinates, 5.68–5.72
Bracketing equations for pocket calculators, 1.31–1.32
Broaching calculations, 5.63–5.66
   pulling forces, 5.64
   pushing forces, 5.64–5.65

Calculations (*see* individual topics)
Calculator techniques, 1.29–1.31
Cams, 8.6–8.18
   calculations for, 8.13, 8.15–8.18
   followers, 8.17
   layout of, 8.7–8.13
Circle, properties of, 2.10
Clamps, tooling, calculations for, 11.9–11.17
Compound angles, calculating, 4.54–4.63
Countersinking, 4.8–4.10
   advance, 4.8–4.9
   calculations, 4.8–4.10

Drill point angles, 5.39–5.42
Drill point advance, 4.8–4.10
Drilling and boring coordinates, 5.67–5.72

Equations, 1.9–1.11
   bracketing, 1.31–1.32
   solving algebraic and trigonometric, 1.7–1.27

External mechanisms
   Geneva mechanism, 11.1–11.3
   ratchets, 8.3–8.4

Fits, classes and calculations for, 12.1–12.9
   common practice tables, 12.2–12.3
   SI fits (ISO), 12.8
   stresses in force fits, 12.9–12.13
   U.S. Customary fits, 12.6

Gears, 7.1–7.15
   formulas for, 7.4–7.12
   bevel, 7.11
   miter, 7.11
   helical, 7.11
   spur, 7.10
   worm, 7.12
Geometric figures, 3.1–3.13
   calculations for areas, volumes, and surfaces of, 2.1–2.10
Geometric constructions, 3.1–3.13
Geometry, principles and laws of, 1.1–1.6

Horsepower requirements
   for milling, 5.27, 5.30
   for turning, 5.5–5.70

Internal mechanisms
   Geneva mechanism, 11.3–11.5
   ratchets, 8.4–8.5

Jig boring coordinates, 5.67–5.72
   calculations for, 5.71–5.72
Jigs and fixtures, clamps for, 11.9–11.17

Linkages
   calculations for, 11.17–11.24
   complex, 11.17, 11.22–11.24
   simple, 11.17–11.21

Mathematical series and uses, 7.4–7.6
Mechanisms, calculations for, 11.1–11.24
  clamping mechanisms, 11.9–11.17
  common mechanisms, 11.1–11.8
  four-bar linkage, 11.17–11.18, 11.21–11.24
  Geneva mechanisms, internal and external, 11.1–11.5
  linkages, 11.17–11.24
  slider-crank, 11.8
Mensuration
  formulas for, 2.1–2.9
  of plane and solid shapes, 2.1–2.9
Metal removal rate (mrr), 5.6, 5.26–5.27
Milling calculations, 5.26–5.38
  angular cuts, 4.54–4.63
  metal removal rate (mrr), 5.6, 5.26–5.27
  notches and V grooves, 4.20–4.22, 4.26–4.31
  tables for, 5.27–5.30, 5.36–5.38
Milling feeds and speeds, 5.26–5.30

Notches, checking, 4.20–4.22, 4.26–4.31
Notching, 4.32–4.38

Open angles, sheet metal, 6.12

Plunge depth calculations for milling notch widths, 4.33–4.36
Punching
  and blanking, 6.32
  sheet metal, forces for, 6.32

Quadratic equations, 1.7

Ratchets, internal and external, 8.1–8.6
  calculations, 8.3–8.6
  geometry of, 8.4, 8.5
  pawls, 8.1–8.2

Sheet metal, 6.1–6.41
  angled corner notching of, 6.28–6.31
  bend radii of, 6.14–6.16
  bending calculations for, 6.8–6.13

Sheet metal (*Cont.*)
  development of, 6.17–6.29
  flat-pattern calculations for, 6.8–6.13
  gauges of, standard, 6.4–6.8
  punching pressures for, 6.32
  shear strengths of, 6.32–6.35
  tooling requirements for, 6.36–6.41
Sine bars, 4.1–4.2
Sine plates, 4.5–4.6
Spade-drilling forces, 5.57–5.61
Spring materials, properties of, 10.16–10.21
Springs, calculations for, 10.1–10.27
  compression, 10.5–10.8
  elastomer, 10.22–10.23
  extension, 10.8–10.11
  flat and beam, 10.14–10.16
  problems with, 10.24–10.26
  torsion, 10.11–10.14
Sprockets, geometry of, 7.15–7.18

Temperature systems, 1.36–1.37
Threads, 5.12–5.20
  calculating pitch diameters of, 9.5–9.6
  measuring pitch diameters of, 9.6–9.12
  pitch diameters of, 9.13–9.15
  pull out calculations for, 9.1–9.5
  tap drills for, 5.47–5.53
  turning, 5.20–5.22
Thread systems, 5.12–5.20
Toggle linkage, 11.18
Tooling clamps, 11.9–11.17
Torque tables, screw and bolt, 9.16
Transposing equations, 1.9–1.11
Trigonometric identities, 1.18–1.21
Trigonometry, 1.11–1.28
  problems, samples of, 1.21–1.27
Turning, 5.1–5.8
  calculations for, 5.1–5.8
  feed tables for, 5.9–5.11
  horsepower requirements for, 5.5–5.6
  metal removal rate (mrr), lathe, 5.6, 5.12
  speed tables for, 5.9–5.11

ABOUT THE AUTHOR

Ronald A. Walsh is one of McGraw-Hill's most successful writers. An electromechanical design engineer for more than 45 years, he wrote *Machining and Metalworking Handbook,* Second Edition, and *Electromechanical Design Handbook,* Third Edition, and is the coauthor of *Engineering Mathematics Handbook,* Fourth Edition, all published by McGraw-Hill. Former director of research and development at the Powercon Corporation and the holder of three U.S. patents, he now consults widely from a base in Upper Marlboro, Maryland.